The Mathematics of Fluid Flow Through Porous Media

The Mathematics of Fluid Flow Through Porous Media

Myron B. Allen
University of Wyoming

Registered Office
John Wiley & Sons, Inc., 111 River Street, Hoboken, NJ 07030, USA

Editorial Office
111 River Street, Hoboken, NJ 07030, USA

For details of our global editorial offices, customer services, and more information about Wiley products visit us at www.wiley.com.

Wiley also publishes its books in a variety of electronic formats and by print-on-demand. Some content that appears in standard print versions of this book may not be available in other formats.

Library of Congress Cataloging-in-Publication Data Applied for:

ISBN: 9781119663843

Cover Design: Wiley
Cover Image: © Myron B. Allen

Set in 9.5/12.5pt STIXTwoText by Straive, Chennai, India

10 9 8 7 6 5 4 3 2 1

To Professor George F. Pinder, who has lit the path for so many.

Contents

Preface

Seldom turns out the way it does in the song.

Robert Hunter

This book provides a semester-length course in the mathematics of fluid flows in porous media. Over a 20-year span, I taught such a course every few years to doctoral students in engineering, mathematics, and geophysics. Most of these students' research involved flow and transport in groundwater aquifers, soils, and petroleum reservoirs. The students' mathematical backgrounds ranged from standard undergraduate engineering requirements to more advanced, graduate-level training.

The book emphasizes analytic aspects of flows in porous media. This focus may seem odd: Most mathematically oriented scholarship in the area is computational in nature, owing both to the heterogeneity of natural porous media and to the inherent nonlinearity of many underground flow models. Nevertheless, while many superb books cover computational methods for flows in porous media, intelligent design of numerical approximations also requires a grasp of certain analytic questions:

- Where do the governing equations come from?
- What physics do they model, and what physics do they neglect?
- What qualitative properties do their solutions exhibit?

Where appropriate, the book discusses numerical implications of these questions.

The exposition should be accessible to anyone who has completed a baccalaureate program in engineering, mathematics, or physics at a US university. The book makes extensive use of multivariable calculus, including the integral theorems of vector field theory, and ordinary differential equations. Several sections exploit concepts from first-semester linear algebra. No prior study of partial differential equations is necessary, but some exposure to them is helpful.

After a brief introduction in Chapter 1, Chapter 2 introduces the mass and momentum balance laws from which the governing partial differential equations arise. This chapter sets the stage for a pattern that appears throughout the book: We derive governing equations, then analyze representative or generic solutions to infer important attributes of the flows.

Chapters 3 through 5 examine models of single-fluid flows, followed by models of the transport of chemical species in the subsurface. After a discussion in Chapter 6 of multiphase flows, traditionally the province of oil reservoir engineers but now also important in groundwater contaminant hydrology and carbon dioxide sequestration, Chapter 7 provides an overview multifluid, multispecies flows, also called compositional flows. This level of complexity admits few analytic solutions. Therefore, Chapter 7 focuses on model formulation.

Two features of the book deserve comment.

- Over 100 exercises, most of them straightforward, appear throughout the text. Their main purpose is to engage the reader in some of the steps required to develop the theory.
- There are four appendices. The first simply lists symbols that have dedicated physical meanings. The remaining appendices cover three common curvilinear coordinate systems, the Buckingham Pi theorem of dimensional analysis, and some aspects of surface integrals. While needed at certain junctures in the text, these topics seem ancillary to the book's main focus.

I owe thanks to dozens of students at the University of Wyoming who endured early versions of the notes for this book. These men and women convinced me of its utility and offered many corrections and suggestions for improvement. Professor Frederico Furtado kindly offered additional corrections, generous encouragement, and insights deeper than he will admit. I also owe sincerest thanks to my colleagues in the University of Wyoming's Department of Mathematics and Statistics, from whom I have learned a lot. I cannot have asked for a better academic home. Finally, my wife, Adele Aldrich, deserves more gratitude than I know how to express, for her support through the entire process.

Laramie, Wyoming   Myron B. Allen
December, 2020

1

Introduction

1.1 Historical Setting

The mathematical theory of fluid flows in porous media has a distinguished history. Most of this theory ultimately rests on Henry Darcy's 1856 engineering study [43], summarized in Section 3.1, of the water supplies in Dijon, France. A year after the publication of this meticulous and seminal work, Jules Dupuit [49], a giant among early groundwater scientists, recognized that Darcy's findings implied a differential equation. This observation proved to be crucial. For the next 75 years or so, the subject grew to encompass problems in multiple space dimensions—hence **partial differential equations** (**PDEs**)—with major contributions emerging mainly from the groundwater hydrology community. Pioneers included Joseph Boussinesq [25, 26], Philipp Forchheimer [53, 54], Charles S. Slichter [136], Edgar Buckingham [30], and Lorenzo A. Richards [129].

Interest in the mathematics of porous-medium flows blossomed as oil production increased in economic importance during the early twentieth century. Prominent in the early petroleum engineering literature in this area are works by P.G. Nutting [110], Morris Muskat and his collaborators [104–107, 159, 160], and Miles C. Leverett and his collaborators [29, 95–97]. Between 1930 and 1960, mathematicians, groundwater hydrologists, petroleum engineers, and geoscientists made tremendous progress in understanding the PDEs that govern underground fluid flows.

Today, mathematical models of porous-medium flow encompass linear and nonlinear PDEs of all major types, as well as systems involving PDEs having different types. The analysis of these equations and their numerical approximations requires an increasing level of mathematical and computational sophistication, and the models themselves have become essential design tools in the management of underground fluid resources.

The Mathematics of Fluid Flow Through Porous Media, First Edition. Myron B. Allen.
© 2021 John Wiley & Sons, Inc. Published 2021 by John Wiley & Sons, Inc.

From a philosophical perspective, credit for these advances belongs to scientists and engineers who clung tenaciously—often in the face of skepticism on the part of more "practically" oriented colleagues—to two premises. The first is that the key to effective modeling resides in careful mathematical reasoning. While this premise seems platitudinous, at any moment in history some practitioners believe that their science is too inherently messy to justify fastidious mathematics. On the contrary, the need for painstaking logical inferences from premises and hypotheses is arguably never greater than when the data are complicated, confusing, or hard to obtain.

The second premise is more subtle: In the absence of good data, sound mathematical models are essential. Far from outstripping the data, mathematical models tell us what data we really need. Moreover, they tell us what qualitative properties we can expect in predictions arising from a given input data set. They also reveal how properties of the data, such as its spatial variability and uncertainty, affect the models' predictive capabilities. If the required data cannot in principle be acquired, if the qualitative properties of the model conflict with the empirical evidence, or if the model cannot, in principle, provide stable predictions in the face of heterogeneity and uncertainty, then we must admit that our understanding is incomplete.

1.2   Partial Differential Equations (PDEs)

Most realistic models of fluid flows in porous media use PDEs, "the natural dialect of continuum science" [62], written at scales appropriate for bench- or field-scale observations. In practical applications, these equations are complicated. They are posed on geometrically irregular, multidimensional domains; they often have highly variable coefficients; they can involve coupled systems of equations; in many applications they are nonlinear. For these reasons, we must often replace the exact PDEs by arithmetic approximations that one can solve using electronic machines.

The practical need for computational methods notwithstanding, a grasp of the analytic aspects of the PDEs remains an important asset for any porous-medium modeler. What types of initial and boundary conditions yield well-posed problems? Do the solutions obey *a priori* bounds based on the initial or boundary data? Do the numerical approximations respect these bounds? Does the PDE tend to smooth or preserve numerically problematic sharp fronts as time advances? Do shocks form from continuous initial data?

In the first half of the twentieth century, pioneering numerical analysts Richard Courant, Kurt Friedrichs, Hans Lewy, and John von Neumann—all immigrants to the United States—recognized that one cannot successfully "arithmetize analysis" [23] without understanding the differential equations. Designing stable, convergent, accurate, and efficient approximations to PDEs requires mathematical insight into the equations being approximated. A visionary 1947 consulting report [152] by von Neumann, developing the first petroleum reservoir simulator designed for a computer, illustrates this principle.

This book aims to promote this type of insight. We examine PDE-based models of porous-medium flows in geometries and settings simple enough to admit analysis without numerical approximations but realistic enough to reveal important structures.

From a mathematical perspective, the study of fluid flows in porous media offers fertile ground for inquiry into PDEs more generally. In particular, this book employs many broadly applicable concepts in the theory of PDEs, including:

1. Mass and momentum balance laws
2. Variational principles
3. Fundamental solutions
4. The principle of superposition
5. Similarity methods
6. Stability analysis
7. The method of characteristics and jump conditions.

Where possible, the narrative introduces these topics in the simplest possible settings before applying them to more complicated problems.

Topic 1, covered in Chapter 2, deserves comment. Few PDE texts at this level discuss balance laws in the detail pursued here. However, it is hard to build intuition about porous-medium flows without knowing the principles from which they arise. The balance laws furnish those principles. On the other hand, a completely rigorous study of balance laws for fluids flowing in porous media would require a monograph-length treatment in its own right. Chapter 2 reflects an attempt to weigh the importance of fundamental principles against the need for a concise explanation of how the governing PDEs emerge from basic laws of physics. The references offer suggestions for deeper inquiry.

We frequently refer to PDEs according to a classification system inherited from the algebra of quadratic equations. The utility of this system becomes more apparent as one becomes more familiar with examples. For now, it suffices to review the system for second-order PDEs in two independent variables having the form

$$a\frac{\partial^2 u}{\partial x^2} + b\frac{\partial^2 u}{\partial x\,\partial y} + c\frac{\partial^2 u}{\partial y^2} = F\left(x, y, u, \frac{\partial u}{\partial x}, \frac{\partial u}{\partial y}\right). \tag{1.1}$$

Here, a, b, and c are functions of the independent variables x and y, which we can replace with x and t in time-dependent problems; $u(x, y)$ is the unknown solution; and F denotes a function of five variables that describes the lower-order terms in the PDE.

The highest-order terms determine the classification. The **discriminant** of Eq. (1.1) is $\Delta = b^2 - 4ac$, which is a function of (x, y). Equation (1.1) is

- **hyperbolic** at any point of the (x, y)-plane where $\Delta(x, y) > 0$;
- **parabolic** at any point of the (x, y)-plane where $\Delta(x, y) = 0$;
- **elliptic** at any point of the (x, y)-plane where $\Delta(x, y) < 0$.

Extending this terminology, we say that a first-order PDE of the form

$$\frac{\partial u}{\partial x} + a \frac{\partial u}{\partial y} = F(x, y, u)$$

is hyperbolic at any point (x, y) where $a(x, y) \neq 0$.

Exercise 1.1 *Verify the following classifications, where c and D are real-valued with D > 0:*

$$\frac{\partial^2 u}{\partial t^2} - c^2 \frac{\partial^2 u}{\partial x^2} = 0 \qquad \text{(one-dimensional wave equation)} \quad \textit{hyperbolic,}$$

$$\frac{\partial u}{\partial t} - D \frac{\partial^2 u}{\partial x^2} = 0 \qquad \text{(one-dimensional heat equation)} \quad \textit{parabolic,}$$

$$\frac{\partial^2 u}{\partial x^2} + \frac{\partial^2 u}{\partial y^2} = 0 \qquad \text{(two-dimensional Laplace equation)} \quad \textit{elliptic.}$$

Mathematicians associate the wave equation with time-dependent processes that exhibit wave-like behavior, the heat equation with time-dependent processes that exhibit diffusive behavior, and the Laplace equation with steady-state processes. These associations arise from applications, some of which this book explores, reinforced by theoretical analyses of the three exemplars in Exercise 1.1. For more information about the classification of PDEs, see [65, Section 2-6].

1.3 Dimensions and Units

In contrast to most texts on pure mathematics, in this book **physical dimensions** play an important role. We adopt the basic physical quantities length, mass, and time, having physical dimensions L, M, and T, respectively. All other physical quantities encountered in this book—except for one case involving temperature in Chapter 7—are derived quantities, having physical dimensions that are products of powers of L, M, and T.

For example, the physical dimension of force F arises from Newton's second law $F = ma$, where m denotes mass and a denotes acceleration:

$$\dim(F) = \dim(ma) = \dim(m) \cdot \dim(a) = M \cdot LT^{-2}.$$

Analyzing the physical dimensions of quantities that arise in physical laws can yield surprisingly powerful mathematical results. Subsequent chapters exploit this concept many times.

Physical laws such as $F = ma$ require a way to assign numerical values to the physical quantities involved. We do this by comparison with standards, a process called **measurement**. For example, to assign a numerical value to the length of an object, we compare it to a length to which we have assigned a numerical value by fiat. A choice of standards for measuring L, M, and T, applied consistently for all occurrences of length, mass, and time, defines a system of **units**. Changing the system of units typically changes the numerical values that we measure, the exception being **dimensionless** quantities, which have dimension 1.

Where practical, this book uses the *Système Internationale* (SI) as the preferred system of units. The current standards for time, length, and mass in the SI are as follows:

- *Time*: One second (s) is the duration of 9 192 631 770 periods of the radiation emitted by the transition between the two hyperfine levels of the ground state of cesium-133. This period of time is approximately 1/86 400 of one Earth day.
- *Length*: One meter (m) is the distance traveled in a vacuum by light in 1/299 792 458 s. This distance is approximately 10^{-7} times the distance from the Earth's geographic north pole to the equator along a great circle.
- *Mass*: One kilogram (kg) is the mass required to fix the value of the Planck constant as $6.62607015 \times 10^{-34}$ kg m^2 s^{-1}, given the definition of one second and 1 m. This mass is approximately that of 10^{-3} m^3 (1 liter) of water at room temperature and pressure.

In some cases, non-SI units are more convenient for measuring physical quantities that arise in the bench- or field-scale study of fluid flows in porous media. When these cases arise, we give the factor that enables conversion to SI units. The fact that scientists and engineers prefer non-SI units in some instances highlights the inherently subjective nature of units: Humans tend to prefer standards that yield numerical values not far from 1 in our everyday experience. One advantage of using dimensionless quantities—a technique employed frequently in this book—is that we avoid this subjectivity.

1.4 Limitations in Scope

Three limitations in scope are worth noting. First, we treat only isothermal flows in porous media, that is, flows at constant temperature. This restriction conveniently allows us to ignore the energy balance equation in deriving governing

PDEs. On the other hand, it also eliminates several types of flows that have important applications, including flows in geothermal reservoirs and thermal methods of enhanced oil recovery, such as steam flooding.

Also glaringly absent from the table of contents is the topic of flows in fractured porous media. Geoscientists correctly point out that most geologic porous media possess fractures, which exert significant influences on fluid flows. Yet the mathematics of flow in fractured porous media remains poorly delineated, owing not so much to the absence of mathematical models (see [21] for a recent overview and [8, 15, 86, 153] for prominent examples) but, more importantly, to the observation that fractures exist at many scales of observation. In some underground formations, one must know something about the geometry of individual fractures to model fluid flows accurately. In these settings, the modeler's challenge is to represent the discrete fracture system (or statistical realizations) on tractably coarse computational grids. In other geologic settings, it suffices to treat the pore network and the fracture network as overlapping porosity systems, and the challenge is to model how fluids move within *and* between them. This spectrum of modeling approaches deserves a monograph of its own.

Also missing from the topics covered here is a discussion of fluid flows in extremely flow-resistant media, often but debatably referred to as nanodarcy flows but more properly characterized as **non-Darcy flows**. Flows of this type have increased in practical importance during the past two decades, owing especially to vastly improved technologies for producing natural gas from shale formations when hydrocarbon commodity prices justify the costs. The physics here are complex, involving gas–rock interactions in interstices whose typical diameters approach the mean free path of the gas molecules. None of the classical macroscopic transport models—such as Darcy's law or Fick's law of diffusion—suffices by itself to capture these phenomena [37, 81]. One can hope that further advances in our understanding of these flows, analogous to the advances described above for classical Darcy flows, will yield more settled mathematical models in years to come.

2

Mechanics

2.1 Kinematics of Simple Continua

At the macroscopic scale of observation, greater than about 10^{-3} m, a natural porous medium such as sandstone is a complex mixture of solids and fluids, separated by interfaces whose geometries are often too small for humans to discern without aid. This book focuses mainly on the macroscopic scale. However, viewed at the microscopic scale, say 10^{-6}–10^{-3} m, the solids and fluids in a porous medium appear as distinct continua, separated by observable interfaces. We begin with the mechanics of these **simple continua**. Section 2.5 extends the discussion to the mechanics of multiconstituent continua, applicable at the macroscopic scale of observation.

The first step is to establish the **kinematics**. This branch of mechanics provides a framework for describing the motions of continua geometrically, without reference to the forces that cause motion. The treatment here is an abbreviated version of material that appears in standard courses on continuum mechanics; for more details consult [4].

2.1.1 Referential and Spatial Coordinates

In continuum mechanics, the term **body** refers to a collection $\mathcal{B}$ of **particles**, sometimes called **material points**. A subset of the body that is a body in its own right is a **part** of the body. We assign to each body a **reference configuration**, which associates with the body a region $\mathcal{R}$ in three-dimensional Euclidean space. In the reference configuration, each particle in the body has a position $\mathbf{X}$, unique to that particle, as shown in Figure 2.1. The vector $\mathbf{X}$ serves as a label, called the **referential** or **Lagrangian** coordinates of the particle. As with a person's home address, from a strictly logical point of view the particle need not ever occupy

The Mathematics of Fluid Flow Through Porous Media, First Edition. Myron B. Allen.
© 2021 John Wiley & Sons, Inc. Published 2021 by John Wiley & Sons, Inc.

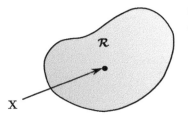

Figure 2.1 A reference configuration of a body, showing the referential coordinates **X** used to label a particle according to its position in the reference configuration.

the point **X**. That said, in some applications it is useful to choose the reference configuration in a way that associates each particle with a position that it occupies at some prescribed time, for example $t = 0$.

The central aim of kinematics is to describe the trajectories of particles, that is, to determine the position **x** in three-dimensional Euclidean space that each particle **X** occupies at every time t. For this purpose we assume that there exists a one-parameter family $\chi(\mathbf{X}, t)$ of vector-valued functions, time t being the parameter, that has the following properties.

1. The vector $\chi(\mathbf{X}, t)$, having dimension L, gives the spatial position **x** of the particle **X** at time t.
2. At each time t, the function $\chi(\cdot, t)$ of the referential coordinates **X** is one-to-one, onto, and continuously differentiable with respect to **X**.
3. Also at each fixed time t, $\chi(\cdot, t)$ has a continuously differentiable inverse χ^{-1} such that $\mathbf{X} = \chi^{-1}(\mathbf{x}, t)$. That is, χ^{-1} tells us which particle **X** occupies the spatial position **x** at time t.
4. For each value of the coordinate **X**, the function $\chi(\mathbf{X}, \cdot)$ is twice continuously differentiable with respect to t.

The function χ is the **deformation** of the body, illustrated in Figure 2.2. We call the vector $\mathbf{x} = \chi(\mathbf{X}, t)$ the **spatial** or **Eulerian** coordinates of the particle **X** at time t.

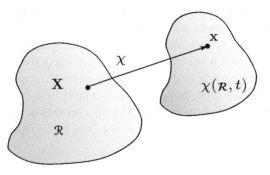

Figure 2.2 The deformation mapping the reference configuration $\mathcal{R}$ onto the body's configuration at time t.

Figure 2.3 Regions $\mathcal{R}$ and $\mathcal{S}$ occupied by a body in two reference configurations, along with the corresponding deformations χ and ψ that map a given particle onto a position vector **x** at time t.

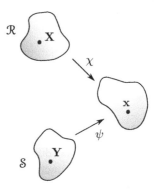

Exercise 2.1 *Let $\mathcal{R}$ and $\mathcal{S}$ be the regions occupied by a body in two different reference configurations, giving the referential coordinates of a certain particle as **X** and **Y**, respectively, as illustrated in Figure 2.3. Let χ and ψ, respectively, denote the deformations associated with these two reference configurations. Thus the spatial position of the particle at time t is $\chi(\mathbf{X}, t) = \mathbf{x} = \psi(\mathbf{Y}, t)$. Justify the relationship $\mathbf{Y} = \psi^{-1}(\chi(\mathbf{X}, t), t)$. This relationship makes it possible to reconcile the analyses of motion by observers who choose different reference configurations.*

2.1.2 Velocity and the Material Derivative

In classical mechanics, it is straightforward to calculate a particle's velocity: Differentiate the particle's spatial position with respect to time. Continuum mechanics employs the same concept. The **velocity** of particle **X** is the time derivative of its position:

$$\frac{\partial \chi}{\partial t}(\mathbf{X}, t). \tag{2.1}$$

This function has dimension LT^{-1}. In taking this partial derivative, we hold the particle **X** fixed and differentiate with respect to t, just as in classical mechanics. We call the velocity (2.1) the **referential velocity** or **Lagrangian velocity**.

We distinguish this velocity from another notion of velocity that arises by measuring what happens at a fixed position in space, as with an anemometer or wind vane attached to a stationary building. This concept of velocity commonly arises in fluid mechanics. In this case, we differentiate with respect to t, holding the spatial coordinate **x** fixed. To calculate this **spatial** or **Eulerian velocity** from the deformation, we first determine which particle $\mathbf{X} = \chi^{-1}(\mathbf{x}, t)$ passes through **x** at time t, then compute the velocity of that particle:

$$\mathbf{v}(\mathbf{x}, t) = \frac{\partial \chi}{\partial t}(\underbrace{\chi^{-1}(\mathbf{x}, t)}_{\mathbf{x}}, t).$$

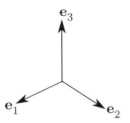

Figure 2.4 Orthonormal basis vectors defining a Cartesian coordinate system.

Since the idea of differentiating with respect to time holding the particle **X** fixed applies to other functions, we adopt a special notation for this operation, called the **material derivative**. If f is a differentiable function of $(\mathbf{X}, t)$—that is, a function of referential coordinates—its material derivative is straightforward:

$$\frac{Df}{Dt}(\mathbf{X}, t) = \frac{\partial f}{\partial t}(\mathbf{X}, t).$$

However, if f is a function of spatial coordinates $(\mathbf{x}, t)$, where $\mathbf{x} = \chi(\mathbf{X}, t)$, calculating its material derivative requires the chain rule. In this context, several common notations for partial differentiation can be ambiguous. If we denote by ∂_1 and ∂_2 the operations of partial differentiation of f with respect to its first and second arguments $\mathbf{x}$ and t, respectively, then

$$\frac{Df}{Dt}(\mathbf{x}, t) = \frac{\partial}{\partial t} f(\chi(\mathbf{X}, t), t)$$

$$= \partial_1(\chi(\mathbf{X}, t), t)\frac{\partial \chi}{\partial t}(\mathbf{X}, t) + \partial_2(\chi(\mathbf{X}, t), t)\frac{\partial t}{\partial t}$$

$$= \nabla f(\mathbf{x}, t) \cdot \mathbf{v}(\mathbf{x}, t) + \frac{\partial}{\partial t} f(\mathbf{x}, t).$$

In the third line of this derivation, ∇f denotes the gradient of the function f, that is, its derivative with respect to the vector-valued spatial position $\mathbf{x}$. With respect to any orthonormal basis $\{\mathbf{e}_1, \mathbf{e}_2, \mathbf{e}_3\}$, as drawn in Figure 2.4,

$$\nabla f = \sum_{i=1}^{3} \frac{\partial f}{\partial x_i} \mathbf{e}_i.$$

In short, for a function f of spatial position and time, the material derivative is

$$\boxed{\frac{Df}{Dt} = \frac{\partial f}{\partial t} + \mathbf{v} \cdot \nabla f.}$$

2.2 Balance Laws for Simple Continua

The partial differential equations (PDEs) governing flows through porous media arise from balance laws. All of the flows treated in this book are isothermal,

Figure 2.5 A time-independent region $\mathcal{V}$ having oriented boundary $\partial\mathcal{V}$ and unit outward normal vector field **n**. The small arrows represent the spatial velocity.

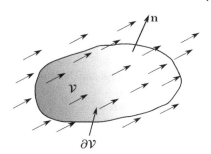

that is, the temperature is constant in time and uniform in space. With this assumption in place, we need only work with the balance laws governing mass and momentum—with one exception noted below.

2.2.1 Mass Balance

Consider first the mass balance. We associate with each body a nonnegative, integrable function $\rho(\mathbf{x}, t)$, called the **mass density**. This function gives the mass contained in any region $\mathcal{V}$ of three-dimensional Euclidean space as the volume integral

$$\int_{\mathcal{V}} \rho(\mathbf{x}, t)\, dv, \tag{2.2}$$

having physical dimension M. Here, dv denotes the element of volume integration. Since ρ is nonnegative, so is the mass. The expression (2.2) requires that $\dim(\rho) = \mathrm{ML}^{-3}$.

The mass balance arises from a simple observation: The rate of change in the mass inside any region $\mathcal{V}$ of three-dimensional space exactly balances the rate of movement of mass across the region's boundary. In symbols,

$$\boxed{\frac{d}{dt}\int_{\mathcal{V}} \rho\, dv = -\int_{\partial\mathcal{V}} \rho\mathbf{v} \cdot \mathbf{n}\, da.} \tag{2.3}$$

This equation is the **integral mass balance**. Here, $\partial\mathcal{V}$ denotes the boundary of $\mathcal{V}$; $\mathbf{n}(\mathbf{x}, t)$ denotes the unit-length vector field orthogonal to $\partial\mathcal{V}$ and pointing outward, as Figure 2.5 depicts; and da denotes the element of surface integration. We call the function $\rho\mathbf{v}$ in the integral on the right side of Eq. (2.3) the **mass flux** per unit area; the integrand $\rho\mathbf{v} \cdot \mathbf{n}$ is the component of mass flux per unit area in the direction of the unit vector **n**, that is, outward from $\mathcal{V}$. The surface integral itself, together with the negative sign, is the **net flux** of mass *inward* across $\partial\mathcal{V}$.

Often of greater utility than the integral equation (2.3) is a pointwise form of the mass balance, valid when the density and velocity are sufficiently smooth. To derive this form, consider a region $\mathcal{V}$ that does not change in time. In this case,

$$\frac{d}{dt} \int_{\mathcal{V}} \rho \, dv = \int_{\mathcal{V}} \frac{\partial \rho}{\partial t} \, dv. \qquad (2.4)$$

Also, by the divergence theorem,

$$-\int_{\partial \mathcal{V}} \rho \mathbf{v} \cdot \mathbf{n} \, da = -\int_{\mathcal{V}} \nabla \cdot (\rho \mathbf{v}) \, dv, \qquad (2.5)$$

where $\nabla \cdot$ denotes the divergence operator. With respect to an orthonormal basis $\{\mathbf{e}_1, \mathbf{e}_2, \mathbf{e}_3\}$,

$$\nabla \cdot (\rho \mathbf{v}) = \sum_{j=1}^{3} \frac{\partial}{\partial x_j} (\rho \, v_j).$$

Applying the identities (2.4) and (2.5) to the integral mass balance (2.3) yields the equivalent equation

$$\int_{\mathcal{V}} \left[\frac{\partial \rho}{\partial t} + \nabla \cdot (\rho \mathbf{v}) \right] dv = 0, \qquad (2.6)$$

valid for any time-independent region $\mathcal{V}$.

If the integrand in Eq. (2.6) is continuous, then the integrand must vanish:

$$\boxed{\frac{\partial \rho}{\partial t} + \nabla \cdot (\rho \mathbf{v}) = 0.} \qquad (2.7)$$

Equation (2.7) is the **differential mass balance**.

Exercise 2.2 *Verify the principle used to derive Eq. (2.7) from the integral equation (2.6). An argument by contradiction may be the easiest approach: Assume that the integrand on the left side of Eq. (2.6) is positive at some point $\mathbf{x}$ at some time t. Since this function is continuous, it must be positive in a neighborhood of $\mathbf{x}$ at time t. Consider a fixed region contained in this neighborhood. A similar argument dispatches the possibility that the integrand is negative at some point.*

Exercise 2.3 *Justify the following equivalent of the mass balance (2.7):*

$$\frac{D\rho}{Dt} + \rho \nabla \cdot \mathbf{v} = 0. \qquad (2.8)$$

The differential mass balance in the form (2.8) facilitates another observation. In certain motions, the density following any particle is constant. In this case, $D\rho/Dt = 0$, so the mass balance implies that

$$\nabla \cdot \mathbf{v} = 0.$$

In this case, we say that the motion is **incompressible**. This concept does not imply anything about the material being modeled; it merely describes the motion based on properties of the velocity field. A compressible material can undergo incompressible motion.

The mass balance is the simplest of the balance laws of continuum mechanics. Other balance laws include the momentum balance, the angular momentum balance, and the energy balance. A related thermodynamic law known, as the entropy inequality, also plays an important role in many settings. In each of these laws, an integral version is fundamental, and it is possible to derive differential versions under certain continuity conditions. For a detailed review of the integral balance laws and the derivation of their differential versions, see [4]. With the exception of several applications of the mass balance discussed in Chapters 5 and 6, the remainder of this book focuses on differential balance laws.

2.2.2 Momentum Balance

The differential momentum balance equation is

$$\rho \frac{D\mathbf{v}}{Dt} - \nabla \cdot \mathsf{T} - \rho\mathbf{b} = \mathbf{0}, \qquad (2.9)$$

often called Cauchy's first law. (For its derivation from an integral form, see [4, Chapter 4]. Strictly speaking, the momentum balance states that there exists a frame of reference in which Cauchy's first law holds.) Each term in Eq. (2.9) is a vector-valued function having dimension $ML^{-2}T^{-2}$. Thus, Cauchy's first law comprises three scalar PDEs.

The terms in Eq. (2.9) require explanation. First, with respect to any orthonormal basis $\{\mathbf{e}_1, \mathbf{e}_2, \mathbf{e}_3\}$,

$$\mathbf{v} \cdot \nabla = \sum_{j=1}^{3} v_j \frac{\partial}{\partial x_j},$$

so applying this operator to $\mathbf{v}$ yields

$$\frac{D\mathbf{v}}{Dt} = \left(\frac{\partial}{\partial t} + \mathbf{v} \cdot \nabla \right)\mathbf{v} = \left(\frac{\partial}{\partial t} + \sum_{j=1}^{3} v_j \frac{\partial}{\partial x_j} \right) \sum_{i=1}^{3} v_i \mathbf{e}_i$$

$$= \sum_{i=1}^{3} \left(\frac{\partial v_i}{\partial t} + \sum_{j=1}^{3} v_j \frac{\partial v_i}{\partial x_j} \right) \mathbf{e}_i,$$

which is clearly a vector-valued function.

Second, the function $\mathbf{b}(\mathbf{x}, t)$ represents the **body force per unit mass**, having dimension LT^{-2}. In this book, the only body force of interest is gravity, and $\mathbf{b}$

reduces to the gravitational acceleration near Earth's surface. The total body force acting on a part of the body is

$$\int_{\mathcal{V}} \rho \mathbf{b} \, dv,$$

where $\mathcal{V}$ is the region occupied by the part.

Third, the function $\mathsf{T}(\mathbf{x}, t)$ is the **stress tensor**. This entity deserves more extended discussion, starting with the term **tensor**. A second-order tensor is a linear transformation that maps vectors into vectors. Its geometric action remains fixed under changes in coordinate systems, a requirement discussed in more detail in Section 3.7. The stress tensor is a linear transformation that describes a type of force different from the body force.

More specifically, any part of a body occupying a region $\mathcal{V}$ in three-dimensional space can experience forces acting on the region's bounding surface $\partial\mathcal{V}$. We account for these forces by introducing **tractions**, having dimension force per unit area:

$$\dim\left(\frac{\text{Force}}{\text{Area}}\right) = \frac{\text{MLT}^{-2}}{\text{L}^2} = \text{ML}^{-1}\text{T}^{-2}.$$

Consider such a region, as drawn in Figure 2.6. At any point where the bounding surface $\partial\mathcal{V}$ is smooth and orientable, there exists an outward pointing unit normal vector $\mathbf{n}$ that is orthogonal to the plane tangent to $\partial\mathcal{V}$ at that point. The stress tensor is a linear transformation T such that the vector field $\mathsf{T}\mathbf{n}$ gives the traction at any point on $\partial\mathcal{V}$. The vector field $\mathsf{T}\mathbf{n}$ need not be collinear with $\mathbf{n}$: The force per unit area acting at a point on $\partial\mathcal{V}$ can have a component tangent to the surface. The total force acting on $\partial\mathcal{V}$ is

$$\int_{\partial\mathcal{V}} \mathsf{T}\mathbf{n} \, da,$$

having dimension MLT^{-2}.

Four additional remarks help clarify the nature of the stress tensor.

1. With respect to any orthonormal basis $\{\mathbf{e}_1, \mathbf{e}_2, \mathbf{e}_3\}$, any linear transformation A has a matrix representation with entries $A_{ij} = \mathbf{e}_i \cdot \mathsf{A}\mathbf{e}_j$. For T, this representation

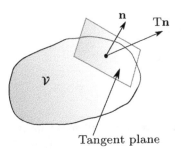

Figure 2.6 A region $\mathcal{V}$ in three-dimensional space with unit outward normal vector field $\mathbf{n}$ and the traction $\mathsf{T}\mathbf{n}$ acting on the boundary $\partial\mathcal{V}$.

has the form

$$\begin{bmatrix} T_{11} & T_{12} & T_{13} \\ T_{21} & T_{22} & T_{23} \\ T_{31} & T_{32} & T_{33} \end{bmatrix}.$$

2. In accordance with Exercise 2.4, with respect to any orthonormal basis, the diagonal entries T_{11}, T_{22}, T_{33} represent forces per unit area acting in directions perpendicular to faces that are orthogonal to $\mathbf{e}_1$, $\mathbf{e}_2$, and $\mathbf{e}_3$, respectively. We refer to these entries as **tensile stresses** when they pull in the same direction as $\mathbf{n}$ and as **compressive stresses** when they push in the opposite direction—namely inward—from $\mathbf{n}$. The off-diagonal entries T_{ij}, where $i \neq j$, are **shear stresses**.

3. A classic theorem in continuum mechanics reduces the angular momentum balance, which we do not discuss here, to the identity $T_{ij} = T_{ji}$ with respect to any orthonormal basis. In other words, the stress tensor is symmetric. See [4, Chapter 4] for details.

4. With respect to an orthonormal basis $\{\mathbf{e}_1, \mathbf{e}_2, \mathbf{e}_3\}$, the divergence $\nabla \cdot \mathsf{T}$ of the tensor-valued function T has the following representation as a vector-valued function:

$$\sum_{j=1}^{3} \begin{bmatrix} \partial T_{j1}/\partial x_j \\ \partial T_{j2}/\partial x_j \\ \partial T_{j3}/\partial x_j \end{bmatrix}.$$

Exercise 2.4 *Consider the action of* T *on each unit basis vector* $\mathbf{e}_i$, $i = 1, 2, 3$, *to examine the forces acting on faces of a cube of material whose edges lie parallel to the Cartesian coordinate axes defined by* $\{\mathbf{e}_1, \mathbf{e}_2, \mathbf{e}_3\}$, *as drawn in Figure 2.7. Justify the assertion that* T_{ij} *represents the ith component of the force per unit area acting on surfaces that lie perpendicular to* $\mathbf{e}_j$.

Figure 2.7 A cube of material illustrating the interpretations of entries of the stress tensor matrix with respect to an orthonormal basis, from [[4], page 109].

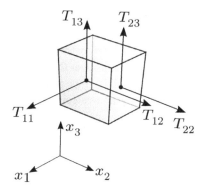

The differential momentum balance (2.9) generalizes Newton's second law of motion. The left side of Eq. (2.9) is proportional to mass × acceleration, while the right side is proportional to a sum of forces. Thus, Eq. (2.9) has the form

$$\frac{1}{\text{Volume}} \left(\text{Mass} \times \text{Acceleration} = \sum \text{Forces} \right).$$

Based on this parallel, fluid mechanicians call

$$\rho \frac{D\mathbf{v}}{Dt} = \rho \frac{\partial \mathbf{v}}{\partial t} + \rho (\mathbf{v} \cdot \nabla)\mathbf{v}$$

the **inertial terms**.

If we view the momentum balance as an equation for the velocity $\mathbf{v}$, the inertial terms make the momentum balance a nonlinear PDE. In many applications to fluid mechanics, this nonlinearity wreaks mathematical havoc. Mercifully, for reasons examined in Chapter 3, the inertial nonlinearity plays a negligible role in the most commonly used models of porous-media flow. However, this observation furnishes scant grounds for complacency. As subsequent chapters demonstrate, other types of nonlinearity play prominent roles in the fluid mechanics of porous media.

2.3 Constitutive Relationships

The mass and momentum balance laws

$$\frac{D\rho}{Dt} + \rho \nabla \cdot \mathbf{v} = 0,$$

$$\rho \frac{D\mathbf{v}}{Dt} - \nabla \cdot \mathsf{T} - \rho \mathbf{b} = \mathbf{0} \tag{2.10}$$

furnish four scalar PDEs involving the 16 scalar functions required to specify ρ, $\mathbf{v}$, T, and $\mathbf{b}$. The symmetry of the stress tensor, $\mathsf{T} = \mathsf{T}^{\mathsf{T}}$, reduces the number of independent scalar functions to 13. From the mathematical point of view, a well posed problem involving Eqs. (2.10) requires $13 - 4 = 9$ additional equations to close the system. We call these equations **constitutive relationships**.

From the engineer's point of view, constitutive relationships define the physical system being modeled. Since the mass and momentum balance laws apply to all materials, by themselves they provide no way to distinguish among different types of fluids and solids. If we regard the differential equations (2.10) as governing the mass density ρ and velocity $\mathbf{v}$, then we need to specify constitutive relationships for the three scalar functions defining the body force $\mathbf{b}$ and the six independent scalar functions $T_{11}, T_{22}, T_{33}, T_{12}, T_{13}, T_{23}$ that define the matrix representation of the stress tensor. This book examines only a small number of constitutive relationships, chosen from the myriad that scientists and engineers have developed to model the remarkable variety of materials found in nature.

Figure 2.8 Coordinate system used to define the depth function $Z(\mathbf{x})$.

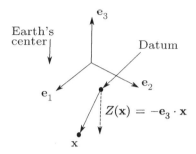

2.3.1 Body Force

For the body force, which is gravity in all of the problems examined here, we adopt the constitutive relationship $\mathbf{b} = -g\,\mathbf{e}_3$. Here $g \simeq 9.8$ m s^{-2} denotes the gravitational acceleration, which varies across Earth's surface, and we adopt a Cartesian coordinate system in which $\mathbf{e}_3$ points away from Earth's center, as shown in Figure 2.8.

An alternative way of writing this expression proves useful in subsequent sections. Define the **depth function** Z as the mapping that assigns to each spatial point $\mathbf{x}$ its depth $Z(\mathbf{x})$ below some datum, at which $Z = 0$, as drawn in Figure 2.8. We often take the datum to be Earth's surface, but other choices are possible. Observe that

$$\nabla Z = \sum_{i=1}^{3} \frac{\partial Z}{\partial x_i} \mathbf{e}_i = -\mathbf{e}_3,$$

which has dimension $\mathsf{LL}^{-1} = 1$. Therefore, we write the constitutive equation for the body force as $\mathbf{b} = g\nabla Z$.

2.3.2 Stress in Fluids

The stress tensor T enjoys a richer set of possibilities. The simplest is the constitutive relationship for an **ideal fluid**, in which $\mathsf{T} = -p\,\mathsf{I}$. Here, $p(\mathbf{x}, t)$ is a scalar function called the **mechanical pressure**, having dimension $\mathsf{ML}^{-1}\mathsf{T}^{-2}$ (force/area). The SI unit for pressure is 1 pascal, abbreviated as 1 Pa and defined as 1 kg m^{-1} s^{-2}. The symbol I denotes the identity tensor. With respect to any orthonormal basis, the stress of an ideal fluid has matrix representation

$$\begin{bmatrix} -p & 0 & 0 \\ 0 & -p & 0 \\ 0 & 0 & -p \end{bmatrix}. \tag{2.11}$$

Thus, in an ideal fluid, there are no shear stresses, and the fluid experiences only compressive and tensile stresses. Also, there are no preferred directions:

$T_{11} = T_{22} = T_{33}$. We describe this fact by saying that the stress tensor is **isotropic**. Section 3.7 discusses isotropic tensors in more detail.

For an ideal fluid in the presence of gravity, the momentum balance reduces to the following equation:

$$\rho \frac{\partial \mathbf{v}}{\partial t} + \rho(\mathbf{v} \cdot \nabla)\mathbf{v} = -\nabla p + \rho g \nabla Z.$$

In problems for which inertial terms are negligible, for example when the fluid is at rest, this equation reduces to

$$\nabla p = \rho g \nabla Z. \tag{2.12}$$

Exercise 2.5 *Integrate the third component of Eq. (2.12) to obtain the **hydrostatic equation**,*

$$p(x_1, x_2, -z) = p(x_1, x_2, 0) + \rho g Z. \tag{2.13}$$

Thus pressure increases linearly with depth in an ideal fluid at rest.

Equation (2.13) closely models the pressure of Earth's atmosphere. At sea level, the pressure of the atmosphere fluctuates around 1.01325×10^5 Pa, which is the definition of a common unit of measurement, 1 atmosphere, abbreviated as 1 atm.

An extension of the ideal fluid stress provides a more realistic constitutive relationship for many fluids. An **incompressible Newtonian fluid** is a material for which

$$\mathsf{T} = -p\mathsf{I} + 2\mu\mathsf{D}. \tag{2.14}$$

Here, D stands for the **stretching tensor**, defined as

$$\mathsf{D} = \frac{1}{2}[\nabla \mathbf{v} + (\nabla \mathbf{v})^\mathsf{T}].$$

With respect to an orthonormal basis, the (i,j)th entry of the matrix representation of $\nabla \mathbf{v}$ is $\partial v_i / \partial x_j$, and $(\nabla \mathbf{v})^\mathsf{T}$ denotes the **transpose** of $\nabla \mathbf{v}$, whose (i,j)th entry is $\partial v_j / \partial x_i$.

The coefficient μ appearing in Eq. (2.14) is the **dynamic viscosity**, a nonnegative function of space and time having dimension $\mathrm{ML^{-1}T^{-1}}$. A common unit for measuring dynamic viscosity is the **centipoise**, abbreviated cP and named after the French physicist Jean Léonard Marie Poiseuille. In SI units, 1 cP = 10^{-3} kg m^{-1} s^{-1}, which is approximately the viscosity of water at a temperature of 20 °C and a pressure of 1 atm. For comparison, the viscosity of air at these conditions is 1.516×10^{-2} cP.

Exercise 2.6 *Find the correct pronunciation of "Poiseuille."*

2.3.3 The Navier–Stokes Equation

Exercise 2.7 *Substitute the constitutive relationship (2.14) into the momentum balance and assume that gravity is negligible (for example, in a shallow horizontal flow) to derive the **Navier–Stokes equation**:*

$$\frac{\partial \mathbf{v}}{\partial t} + (\mathbf{v} \cdot \nabla)\mathbf{v} = -\frac{1}{\rho}\nabla p + \nu \nabla^2 \mathbf{v}. \tag{2.15}$$

*Here, $\nu = \mu/\rho$ is the **kinematic viscosity**, having dimension $L^2 T^{-1}$, and $\nabla^2 \mathbf{v}$ has the following representation with respect to an orthonormal basis:*

$$\sum_{i=1}^{3} \begin{bmatrix} \partial^2 v_1 / \partial x_i^2 \\ \partial^2 v_2 / \partial x_i^2 \\ \partial^2 v_3 / \partial x_i^2 \end{bmatrix}.$$

Sir George Gabriel Stokes was an Irish-born Cambridge professor who made extraordinary contributions to mathematical physics. Claude-Louis Navier was a French mechanical engineer and professor of mathematics in the early nineteenth century.

Exercise 2.8 *Find the correct pronunciation of "Navier."*

Owing largely to mathematical difficulties associated with the inertial terms, the Navier–Stokes equation remains a source of some of the most refractory unsolved problems in mathematics. Proving the existence and smoothness of solutions under general conditions remains one of six unsolved Millennial Prize Problems identified in 2000 by the Clay Institute for Mathematics [79].

To gauge the importance of inertial effects in specific problems, it is useful to cast Eq. (2.15) in terms of dimensionless variables—that is, variables having physical dimension 1. This technique filters out subjective effects associated with the analyst's choice of measurement units, mentioned in Section 1.3.

For concreteness, consider the flow of an incompressible Newtonian fluid in an infinite spatial domain surrounding a solid sphere having radius R, as drawn in Figure 2.9. We examine a simplified version of this flow, called the **Stokes problem**, in Section 2.4. Assume that, as distance from the sphere increases, $\mathbf{v} \to v_\infty \mathbf{e}_1$. Using the radius R and the far-field fluid speed v_∞ as scaling parameters, define the following dimensionless variables:

$$\xi = \frac{\mathbf{x}}{R}, \quad \tau = \frac{v_\infty t}{R}, \quad \mathbf{v}^* = \frac{\mathbf{v}}{v_\infty}, \quad p^* = \frac{p}{\rho v_\infty^2}.$$

By the chain rule, for any sufficiently differentiable function φ,

$$\nabla \varphi = \sum_{i=1}^{3} \frac{\partial \varphi}{\partial x_i} \mathbf{e}_i = \sum_{i=1}^{3} \frac{d\xi_i}{dx_i} \frac{\partial \varphi}{\partial \xi_i} \mathbf{e}_i = \frac{1}{R} \nabla_\xi \varphi,$$

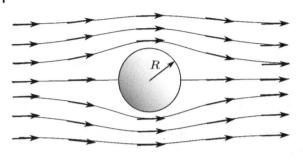

$$\nabla^2 \varphi = \sum_{i=1}^{3} \frac{\partial^2 \varphi}{\partial x_i^2} = \frac{1}{R^2} \sum_{i=1}^{3} \frac{\partial^2 \varphi}{\partial \varsigma_i^2} = \frac{1}{R^2} \nabla_\xi^2 \varphi,$$

$$\frac{\partial \varphi}{\partial t} = \frac{d\tau}{dt} \frac{\partial \varphi}{\partial t} = \frac{v_\infty}{R} \frac{\partial \varphi}{\partial \tau}.$$

Here,

$$\nabla_\xi = \sum_{i=1}^{3} \mathbf{e}_i \frac{\partial}{\partial \varsigma_i}$$

denotes the gradient operator with respect to the dimensionless spatial variable ξ.

Exercise 2.9 *Substitute these operators into the Navier–Stokes equation (2.15) and simplify to get the dimensionless Navier–Stokes equation:*

$$\frac{\partial \mathbf{v}^*}{\partial \tau} + (\mathbf{v}^* \cdot \nabla) \mathbf{v}^* = -\nabla_\xi p^* + \frac{1}{\mathrm{Re}} \nabla_\xi^2 \mathbf{v}^*, \tag{2.16}$$

where $\mathrm{Re} = R v_\infty / \nu.$

The dimensionless parameter Re in Eq. (2.16) is the **Reynolds number**, named after Irish-born fluid mechanician Osborne Reynolds [128]. This number serves as a unit-free gauge of the ratio of inertial effects to viscous effects and, heuristically, as an index of mathematical intractability. We associate the regime Re < 1 with slow flows in which viscous effects dominate those associated with inertia. When Re is much smaller than 1, it is common to neglect the inertial terms.

2.4 Two Classic Problems in Fluid Mechanics

As mentioned in Section 2.3, the Navier–Stokes equation (2.15) poses formidable mathematical challenges. Exact solutions are known only in special geometries and only under highly restrictive assumptions, many of which allow us to neglect the nonlinear inertial term $(\mathbf{v} \cdot \nabla)\mathbf{v}$. We now examine two such problems in fluid mechanics that bear on the analysis of flows in porous media. Each serves as a highly simplified model of fluid flow in the interstices of a porous medium, and each involves significant reductions in complexity compared with the full Navier–Stokes equation. The **Hagen–Poiseuille problem** is a simple model of fluid flows in a straight, cylindrical tube, which one can envision as an idealized pore channel. The **Stokes problem** models slow, viscous flows around a solid sphere, which one can imagine as an idealized solid grain.

2.4.1 Hagen–Poiseuille Flow

One of the earliest known exact solutions to the Navier–Stokes equation arose from a simple but important model examined by Gotthilf Hagen, a German fluid mechanician, and French physicist J.L.M. Poiseuille, mentioned in Section 2.3. Citing Hagen's 1839 work [67], in 1840, Poiseuille [122] developed a classic solution for flow through a pipe. The derivation presented here follows that given by British mathematician G.K. Batchelor [[16], Section 4.2].

Consider steady flow in a thin, horizontal, cylindrical tube having circular cross-section and radius R. Let the fluid's density and viscosity be constant. Orient the Cartesian coordinate system so that the x_1-axis coincides with the axis of the tube.

The problem simplifies if we temporarily convert to **cylindrical coordinates**, defined by the coordinate transformation

$$\Psi \left(\begin{bmatrix} z \\ r \\ \theta \end{bmatrix} \right) = \begin{bmatrix} z \\ r \cos \theta \\ r \sin \theta \end{bmatrix} = \begin{bmatrix} x_1 \\ x_2 \\ x_3 \end{bmatrix}, \tag{B.5}$$

reviewed in Appendix B. Here z represents position along the axis of the tube, r represents distance from the axis, and the angle θ represents the azimuth about the axis. In this coordinate system, the Laplace operator has the form

$$\nabla^2 = \nabla \cdot \nabla = \frac{\partial^2}{\partial z^2} + \frac{1}{r} \frac{\partial}{\partial r} \left(r \frac{\partial}{\partial r} \right) + \frac{1}{r^2} \frac{\partial^2}{\partial \theta^2}. \tag{B.7}$$

Appendix B reviews the derivation of this expression.

In view of the symmetry of the problem about the axis of the tube, we seek solutions of the form

$$\mathbf{v}(z, r, \theta) = (v(r), 0, 0), \tag{2.17}$$

that is, the axial component depends only on distance from the axis of the tube, and the radial and azimuthal coordinates of the velocity vanish, as drawn in Figure 2.10. We allow the pressure to vary with z.

Exercise 2.10 *Show that, under these conditions, the nonlinear term $(\mathbf{v} \cdot \nabla)\mathbf{v}$ vanishes. Work in Cartesian coordinates.*

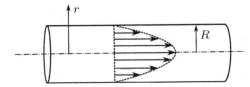

Figure 2.10 Profile of flow through a thin circular cylinder having radius R.

Since the flow is steady, the Navier–Stokes equation therefore reduces to

$$0 = -\nabla p + \frac{1}{\mathrm{Re}} \nabla^2 \mathbf{v}.$$

(Here we use the dimensionless form.)

For a fluid velocity of the form (2.17), we need to only solve the first coordinate equation,

$$0 = -p'(z) + \frac{1}{\mathrm{Re}} \frac{1}{r} \frac{d}{dr} \left(r \frac{dv}{dr} \right). \tag{2.18}$$

Since the second term on the right side of Eq. (2.18) is independent of z by Eq. (2.17), so is the pressure gradient p'. It follows that p' is constant, and hence p varies linearly along the tube. Equation (2.18) therefore reduces to the following ordinary differential equation:

$$\frac{1}{r} \frac{d}{dr} \left(r \frac{dv}{dr} \right) = p' \, \mathrm{Re}.$$

Exercise 2.11 *Verify that the general solution to this equation has the form*

$$v = \frac{p' \, \mathrm{Re}}{4} \left(r^2 + C_1 \log r + C_2 \right),$$

where log *stands for the natural logarithm and C_1 and C_2 denote arbitrary constants.*

For boundary conditions, we assume no slip at the wall of the tube and insist that the velocity along the axis of the tube remain finite:

$$v(R) = 0, \tag{2.19}$$

$$\lim_{r \to 0} |v(r)| < \infty. \tag{2.20}$$

The condition (2.20) requires that $C_1 = 0$.

Exercise 2.12 *Impose the no-slip boundary condition (2.19) to show that*

$$v(r) = \frac{p' \, \mathrm{Re}}{4} \left(R^2 - r^2 \right). \tag{2.21}$$

Equation (2.21) indicates that the fluid velocity has a parabolic profile, with the velocity reaching its maximum magnitude along the axis of the tube and vanishing at the walls. We encounter this profile again in Section 5.1.2, in discussing why solutes spread as they move through the microscopic channels of a porous medium.

2.4.2 The Stokes Problem

Another classic problem derived from the Navier–Stokes equation examines the slow, incompressible, viscous flow of a fluid around a solid sphere of radius R, as drawn in Figure 2.9. In the case of steady flow when the Reynolds number is much smaller than 1, we neglect the inertial terms, arriving at the following mass and momentum balance equations:

$$\nabla \cdot \mathbf{v} = 0,$$

$$\mu \nabla^2 \mathbf{v} = \nabla p.$$

On the surface of the solid sphere, the velocity vanishes, while as one moves far away from the sphere the velocity approaches a uniform far-field value:

$$\mathbf{v}(\mathbf{x}) = \mathbf{0}, \qquad\qquad \|\mathbf{x}\| = R;$$

$$\mathbf{v}(\mathbf{x}) \to v_\infty \mathbf{e}_1, \qquad\qquad \|\mathbf{x}\| \to \infty.$$

In 1851, in a *tour de force* of vector calculus, Stokes [139] published the solution to this boundary-value problem, along with an expression for the total viscous force exerted on the sphere: $\mathbf{F} = F\mathbf{e}_1 = 6\pi\mu R v_\infty \mathbf{e}_1$. This force is called the **Stokes drag**.

For our purposes, we need not examine the calculation of F in detail. Instead, we use a simpler **dimensional analysis**, exploiting concepts from elementary linear algebra, to deduce the functional form of the drag force. Since the only parameters in the boundary-value problem are μ, v_∞, and R, any solution to the problem of calculating F defines a relationship of the form

$$\varphi(F, \mu, v_\infty, R) = 0, \tag{2.22}$$

for some function φ. By a theorem widely attributed to American physicist Edgar Buckingham [31], this relationship, involving variables that have physical dimensions, implies the existence of an equivalent relationship

$$\Phi(\Pi_1, \Pi_2, \dots) = 0$$

involving only dimensionless variables $\Pi_1, \Pi_2, \dots$ Appendix C reviews this theorem.

Thus, we seek relationship equivalent to Eq. (2.22), involving only dimensionless variables formed using powers of the dimensional variables F, μ, v_∞, and R.

For any such variable, denoted generically by Π,

$$
\begin{aligned}
1 = \dim(\Pi) &= \dim(F^{n_1}\mu^{n_2}v_\infty^{n_3}R^{n_4}) \\
&= (MLT^{-2})^{n_1}(ML^{-1}T^{-1})^{n_2}(LT^{-1})^{n_3}L^{n_4} \\
&= M^{n_1+n_2}L^{n_1-n_2+n_3+n_4}T^{-2n_1-n_2-n_3},
\end{aligned}
\tag{2.23}
$$

for exponents n_1, n_2, n_3, n_4 to be determined. Equation (2.23) implies that the exponents of M, L, and T must vanish, yielding the following homogeneous linear system for $n_1, n_2, n_3,$ and n_4:

$$
\begin{bmatrix}
1 & 1 & 0 & 0 \\
1 & -1 & 1 & 1 \\
-2 & -1 & -1 & 0
\end{bmatrix}
\begin{bmatrix}
n_1 \\ n_2 \\ n_3 \\ n_4
\end{bmatrix}
=
\begin{bmatrix}
0 \\ 0 \\ 0
\end{bmatrix}.
\tag{2.24}
$$

Exercise 2.13 *Row-reduce Eq. (2.24) to deduce that n_4 is a free variable, for which one may choose any value, and $n_3 = n_2 = -n_1 = n_4$.*

Arbitrarily picking $n_4 = -1$ yields the single dimensionless variable $\Pi = F\mu^{-1}v_\infty^{-1}R^{-1}$; all other dimensionless variables for this problem must be multiples of this product.

The calculation in Exercise 2.13 shows that any relationship equivalent to Eq. (2.22) but involving only dimensionless variables has the form $\Phi(\Pi) = 0$. Solutions to such an equation are constant values of Π. Setting $\Pi = C$ for a generic constant C, we conclude that Stokes drag has the form

$$
F = C\mu R v_\infty.
\tag{2.25}
$$

This result is consistent with that of Stokes's original calculation, except that we have an undetermined constant C instead of 6π.

To anticipate the constitutive theory of flows in porous media, discussed in Chapter 3, observe that the drag on the solid particle in Eq. (2.25) is proportional to the fluid velocity and the fluid viscosity, and it involves a geometric factor R. This result suffices for the derivation pursued in Section 3.1.

2.5 Multiconstituent Continua

The mechanics discussed so far cannot distinguish among the various solid and fluid bodies that make up a porous medium. To accommodate mixtures of different types of materials, such as the solid and fluid in a porous medium, we must adopt additional physics.

2.5.1 Constituents

The first step in extending the mechanics of single continua is to consider a set of bodies B_α, $\alpha = 1, 2, \ldots, N$, called **constituents**. For example, in a porous medium, rock and water can be constituents. We postulate that each spatial point $\mathbf{x}$ can be occupied by particles from every constituent. In this sense, the bodies $B_1, B_2, \ldots, B_N$ constitute overlapping continua. This postulate clearly fails at scales of observation at which the constituents appear to occupy distinct regions of space. But for many natural porous media found in Earth's subsurface, the postulate yields reasonable models at scales of observation greater than about 10^{-3} m.

Paralleling the development for single continua, for each constituent B_α, we fix a reference configuration that assigns, to each particle in B_α, a point $\mathbf{X}_\alpha$ in three-dimensional space. The vector $\mathbf{X}_\alpha$ serves as a label for the particle. We denote by $\mathcal{R}_\alpha$ the region in three-dimensional Euclidean space occupied by all of these vectors for the constituent B_α.

We also associate with each constituent B_α a one-parameter family $\chi_\alpha(\cdot, t)$ of mappings from $\mathcal{R}_\alpha$ to three-dimensional Euclidean space such that:

1. The vector $\mathbf{x} = \chi_\alpha(\mathbf{X}_\alpha, t)$, having dimension L, gives the spatial position of the particle $\mathbf{X}_\alpha$ at time t, as illustrated in Figure 2.11.
2. At each time t, the function $\chi_\alpha(\cdot, t)$ of the coordinate $\mathbf{X}_\alpha$ is one-to-one, onto, and continuously differentiable with respect to $\mathbf{X}_\alpha$.
3. Also at each time t, $\chi_\alpha(\cdot, t)$ has continuously differentiable inverse χ_α^{-1} such that $\mathbf{X}_\alpha = \chi_\alpha^{-1}(\mathbf{x}, t)$. That is, χ_α^{-1} tells us which particle from constituent B_α occupies the spatial position $\mathbf{x}$ at time t.
4. For each value of the coordinate $\mathbf{X}_\alpha$, the function $\chi_\alpha(\mathbf{X}_\alpha, \cdot)$ is twice continuously differentiable with respect to t.

We call χ_α the **deformation** of constituent B_α.

Figure 2.11 A reference configuration and the deformation at times t_1 and t_2 for constituent α in a multiconstituent continuum.

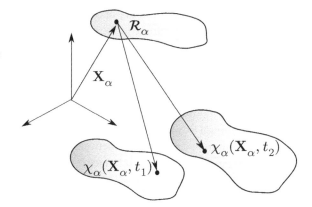

As in the single-continuum case, the **referential** or **Lagrangian velocity** of B_α is

$$\frac{\partial \chi_\alpha}{\partial t}(\mathbf{X}_\alpha, t).$$

To find the velocity of constituent α at a fixed spatial point $\mathbf{x}$ at time t, we first find the particle $\mathbf{X}_\alpha = \chi_\alpha^{-1}(\mathbf{x}, t)$ that occupies $\mathbf{x}$ at time t, then compute the **spatial** or **Eulerian velocity**:

$$\mathbf{v}_\alpha(\mathbf{x}, t) = \frac{\partial \chi_\alpha}{\partial t}\underbrace{(\chi_\alpha^{-1}(\mathbf{x}, t), t)}_{\mathbf{X}_\alpha}.$$

We associate with each constituent B_α a **material derivative**, which gives the time rate of change following a fixed particle $\mathbf{X}_\alpha$. For functions of $(\mathbf{X}_\alpha, t)$, the material derivative is simply the partial derivative with respect to t:

$$\frac{D^\alpha f}{Dt}(\mathbf{X}_\alpha, t) = \frac{\partial f}{\partial t}(\mathbf{X}_\alpha, t).$$

For functions of $(\mathbf{x}, t)$, an application of the chain rule similar to that employed in Section 2.1 for simple continua yields

$$\frac{D^\alpha f}{Dt}(\mathbf{x}, t) = \frac{\partial f}{\partial t}(\mathbf{x}, t) + \mathbf{v}_\alpha(\mathbf{x}, t) \cdot \nabla f(\mathbf{x}, t).$$

2.5.2 Densities and Volume Fractions

As with single continua, we assign to each constituent B_α a **mass density** $\rho_\alpha(\mathbf{x}, t)$ such that the mass of the constituent in any measurable region $\mathcal{V}$ of three-dimensional space at time t is

$$\int_{\mathcal{V}} \rho_\alpha(\mathbf{x}, t) \, dv.$$

Engineers call ρ_α the **bulk density** of constituent B_α; it gives the mass of the constituent per unit of total volume in the continuum.

In the context of porous media, this last observation prompts a discussion of two different categories of multiconstituent continua. The first, which we call **multiphase continua** or **immiscible continua**, includes materials for which microscopic observation reveals continuum-scale interfaces that affect the mechanics. Figure 2.12 illustrates the idea schematically. An example of this type of continuum is water-saturated sandstone. In this porous medium, there exists a continuum-scale interface between the rock and the fluid, but in most sandstones, the geometry of the interface is observable only at scales smaller than about 10^{-3} m.

One way to think of this type of continuum is to regard macroscopic observation as a spatial averaging process. In this view, at each point in space we replace

Figure 2.12 Sketch of a fluid-saturated porous medium showing three possible representative elementary volumes.

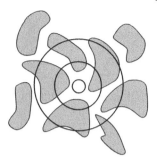

detailed properties of the material with properties averaged over a **representative elementary volume** (REV) having a characteristic radius, as Figure 2.12 illustrates [121]. In the case of water-saturated sandstone, if the radius of the REV ranges over values comparable to typical rock-grain diameters, then the fraction of the REV occupied by fluid oscillates as the radius increases. The oscillation arises because, over this range of radii, the inclusion or exclusion of individual grains results in significant changes in the value of the average.

For the concept of a multiconstituent continuum to furnish a useful model of the porous medium, there must exist a range of REV radii—typically exceeding several rock-grain diameters—in which the fraction of the REV occupied by fluid exhibits a stable value, as drawn in Figure 2.13. Henceforth, we assume that the porous medium possesses a range of REV radii satisfying this condition. We also assume that this range includes radii that are small compared with the macroscopic scale of observation, so that it is reasonable to model the porous medium as a set of overlapping continua.

Under this assumption, we assign to each constituent B_α a **volume fraction** $\phi_\alpha(\mathbf{x}, t)$. This function gives the fraction of any region $\mathcal{V}$ of three-dimensional space occupied by material from the constituent as

$$\int_{\mathcal{V}} \phi_\alpha(\mathbf{x}, t) \, dv.$$

Figure 2.13 Conceptual plot of REV-averaged volume fraction versus radius of averaging window, showing how averaged values can stabilize for a range of averaging radii.

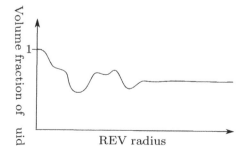

If we account for all of the volume in the continuum, then the volume fractions obey the constraint

$$\sum_{\alpha=1}^{N} \phi_{\alpha} = 1.$$

In this case, we define the **true density** of constituent B_{α} as

$$\gamma_{\alpha} = \frac{\rho_{\alpha}}{\phi_{\alpha}}.$$

The bulk density ρ_{α} has dimension (mass of α)/(volume of continuum), while the true density γ_{α} has dimension (mass of α)/(volume of α).

In the second category of multiconstituent continua, segregation of constituents is observable only at molecular length scales, so continuum-scale interfaces between the constituents do not exist. Saltwater is an example: The particles of Na^+, Cl^-, and H_2O are segregated at length scales of roughly 10^{-10} m, far smaller than the continuum scale. For such **multispecies** or **miscible** multiconstituent continua, the concept of a continuum-scale volume fraction does not apply.

With this framework in place, we define several functions associated with the continuum. The **mixture density** is

$$\rho = \sum_{\alpha=1}^{N} \rho_{\alpha},$$

which we can write for multiphase continua as follows:

$$\rho = \sum_{\alpha=1}^{N} \phi_{\alpha} \gamma_{\alpha}.$$

The mass-weighted or **barycentric velocity** is

$$\mathbf{v} = \frac{1}{\rho} \sum_{\alpha=1}^{N} \rho_{\alpha} \mathbf{v}_{\alpha}.$$

Sometimes it is useful to refer to the **barycentric derivative**, which for a differentiable function $f(\mathbf{x}, t)$ has the form

$$\frac{Df}{Dt}(\mathbf{x}, t) = \frac{\partial f}{\partial t}(\mathbf{x}, t) + \mathbf{v}(\mathbf{x}, t) \cdot \nabla f(\mathbf{x}, t). \tag{2.26}$$

Finally, the **diffusion velocity** of constituent α is

$$v_{\alpha} = \mathbf{v}_{\alpha} - \mathbf{v}. \tag{2.27}$$

Exercise 2.14 *Show that*

$$\sum_{\alpha=1}^{N} \rho_{\alpha} v_{\alpha} = \mathbf{0}.$$

2.5.3 Multiconstituent Mass Balance

The balance laws for single continua extend to multiconstituent continua in a manner that allows for exchanges of mass, momentum, and other conserved quantities among the constituents.

For the differential mass balance, the extension has the following form:

$$\sum_{\alpha=1}^{N} \left(\frac{D^\alpha \rho_\alpha}{Dt} + \rho_\alpha \nabla \cdot \mathbf{v}_\alpha \right) = 0. \tag{2.28}$$

To see how this equation allows for exchanges of mass among constituents, rewrite it as follows:

$$\frac{D^\alpha \rho_\alpha}{Dt} + \rho_\alpha \nabla \cdot \mathbf{v}_\alpha = r_\alpha, \quad \alpha = 1, 2, \dots, N, \tag{2.29}$$

where

$$\sum_{\alpha=1}^{N} r_\alpha = 0. \tag{2.30}$$

Mathematically, this new form amounts to a trivial reformulation. Physically, it captures the exchange of mass into each constituent B_α from other constituents, at a rate given by the **mass exchange rate** r_α, having dimension $ML^{-3}T^{-1}$. Mass exchange can occur via several mechanisms:

- Phase changes, such as melting, freezing, evaporation, and condensation;
- Interphase mass transfer, such as dissolution or adsorption;
- Chemical reactions, which transform molecular species into different molecular species.

For multiphase continua, Eq. (2.29) has an equivalent form:

$$\frac{D^\alpha}{Dt}(\phi_\alpha \gamma_\alpha) + \phi_\alpha \gamma_\alpha \nabla \cdot \mathbf{v}_\alpha = r_\alpha, \quad \alpha = 1, 2, \dots, N$$

again subject to the constraint (2.30).

It is common to write the multiconstituent mass balance in terms of constituent **mass fractions**, defined as $\omega_\alpha = \rho_\alpha / \rho$ and having dimension (mass of α)/(total mass). Doing so yields the following equivalent forms for the mass balance equation for each constituent α, all subject to the constraint (2.30):

$$\frac{D^\alpha}{Dt}(\rho \omega_\alpha) + \rho \omega_\alpha \nabla \cdot \mathbf{v}_\alpha = r_\alpha, \qquad \alpha = 1, 2, \dots, N;$$

$$\frac{\partial}{\partial t}(\rho \omega_\alpha) + \nabla \cdot (\rho \omega_\alpha \mathbf{v}_\alpha) = r_\alpha, \qquad \alpha = 1, 2, \dots, N;$$

$$\underbrace{\frac{\partial}{\partial t}(\rho \omega_\alpha)}_{(I)} + \underbrace{\nabla \cdot (\rho \omega_\alpha \mathbf{v})}_{(II)} + \underbrace{\nabla \cdot \mathbf{j}_\alpha}_{(III)} = \underbrace{r_\alpha}_{(IV)}, \qquad \alpha = 1, 2, \dots, N;$$

where $\mathbf{j}_\alpha = \rho \omega_\alpha \mathbf{v}_\alpha$ is the **diffusive flux** of constituent α. In the last form, we refer to the terms labeled (I), (II), (III), and (IV) as the **accumulation**, **advection**, **diffusion**, and **reaction** terms, respectively.

The following exercise reassuringly shows that the multiconstituent mass balance reduces to the single-constituent mass balance if we use the definitions of the mixture density ρ and the barycentric velocity $\mathbf{v}$ and ignore the distinctions among constituents.

Exercise 2.15 *Use the definitions of the multiconstituent density ρ and the barycentric velocity $\mathbf{v}$ to show that Eq. (2.28) is equivalent to*

$$\frac{D\rho}{Dt} + \rho \nabla \cdot \mathbf{v} = 0.$$

2.5.4 Multiconstituent Momentum Balance

The differential momentum balance for multicomponent continua, in a form paralleling Eqs. (2.29) and (2.30), is

$$\rho_\alpha \frac{D^\alpha \mathbf{v}_\alpha}{Dt} - \nabla \cdot \mathbf{T}_\alpha - \rho_\alpha \mathbf{b}_\alpha = \mathbf{m}_\alpha - \mathbf{v}_\alpha r_\alpha, \quad \alpha = 1, 2, \ldots, N; \tag{2.31}$$

$$\sum_{\alpha=1}^{N} \mathbf{m}_\alpha = \mathbf{0}. \tag{2.32}$$

Here, $\mathbf{m}_\alpha$ represents the rate of momentum exchange into α from other constituents, excluding momentum exchanges associated purely with the transfer of mass into α from other constituents. The term $-\mathbf{v}_\alpha r_\alpha$ gives the rate of momentum exchange into α attributable to mass exchange from other constituents. Equation (2.31) plays a central role in modeling fluid velocities in porous media, as discussed in Sections 3.1 and 3.2.

As with the multiconstituent mass balance equation, one can retrieve the momentum balance for a simple continuum by summing over all constituents and ignoring the distinction among them. This derivation requires a bit of tensor notation encountered again in Section 5.1.

Exercise 2.16 *For any two vectors $\mathbf{a}$, $\mathbf{b}$, the **dyadic product** $\mathbf{a} \otimes \mathbf{b}$ is a tensor having the following action on any vector $\mathbf{u}$:*

$$(\mathbf{a} \otimes \mathbf{b}) \mathbf{u} = \mathbf{a}(\mathbf{b} \cdot \mathbf{u}). \tag{2.33}$$

Verify that the mapping $\mathbf{u} \mapsto \mathbf{a}(\mathbf{b} \cdot \mathbf{u})$ is linear.

Exercise 2.17 *Recall from Section 2.2 that the matrix representation of any tensor A with respect to an orthonormal basis $\{\mathbf{e}_1, \mathbf{e}_2, \mathbf{e}_3\}$ has entries $\mathbf{e}_i \cdot A\mathbf{e}_j$. Compute the matrix representation of $\mathbf{a} \otimes \mathbf{b}$.*

Exercise 2.18 *Sum Eq. (2.31) and use Eq. (2.32), together with the definitions of multiconstituent density ρ and barycentric velocity $\mathbf{v}$, to get*

$$\rho \frac{D\mathbf{v}}{Dt} - \nabla \cdot \mathsf{T} - \rho \mathbf{b} = 0,$$

where

$$\mathbf{b} = \frac{1}{\rho} \sum_{\alpha=1}^{N} \rho_\alpha \mathbf{b}_\alpha$$

gives the total body force per unit mass and

$$\mathsf{T} = \sum_{\alpha=1}^{N} (\mathsf{T}_\alpha - \rho \mathbf{v}_\alpha \otimes \mathbf{v}_\alpha). \qquad (2.34)$$

The tensor T defined in Eq. (2.34) contains an anticipated part,

$$\mathsf{T}_I = \sum_{\alpha=1}^{N} \mathsf{T}_\alpha,$$

called the **inner stress**, and a contribution arising from diffusion velocities,

$$\mathsf{T}_R = -\rho \sum_{\alpha=1}^{N} \mathbf{v}_\alpha \otimes \mathbf{v}_\alpha,$$

sometimes called the **Reynolds stress**, a term borrowed from the theory of turbulence.

3

Single-fluid Flow Equations

We owe the earliest mathematical model of fluid flow in porous media to French engineer Henry Darcy, who in 1856 published a lengthy report [43] of his investigations into a filtration-based water supply system for Dijon, France. From Darcy's empirical observations arose the flow law that bears his name. Darcy's findings have given rise to a rich technical literature since publication of his work. This chapter explores a systematic derivation of Darcy's law from principles developed in Chapter 2, then examines several mathematical aspects of the resulting flow equations.

3.1 Darcy's Law

Figure 3.1 illustrates Darcy's filtration apparatus. By comparing the volumetric flow rate Q of water at the outlet of a sand column with the heights h_1 and h_2 of fluid in manometers located at the top and bottom, respectively, of the column, he found that

$$Q = KA\frac{h_1 - h_2}{\ell}. \tag{3.1}$$

Here, ℓ and A denote the length and cross-sectional area of the column, having dimensions L and L^2, respectively. Q has dimension $L^3 T^{-1}$; and K, having dimension LT^{-1}, stands for a positive constant that depends on the sand used in the column.

Soon after Darcy published his study, French engineer Jules Dupuit [49] used Eq. (3.1) in a differential form, letting $\ell \to 0$ and combining the result with a mass balance equation to model water flow near wells in confined and unconfined aquifers. Later, Austrian engineer Philipp Forchheimer [53] and American mathematician Charles S. Slichter [136, p. 330 ff.] were among the early investigators to

The Mathematics of Fluid Flow Through Porous Media, First Edition. Myron B. Allen.
© 2021 John Wiley & Sons, Inc. Published 2021 by John Wiley & Sons, Inc.

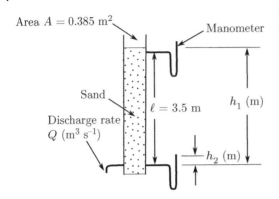

Area $A = 0.385$ m^2

Manometer

Sand

$\ell = 3.5$ m

Discharge rate
Q (m^3 s^{-1})

h_1 (m)

h_2 (m)

Figure 3.1 Schematic diagram of Darcy's apparatus for measuring water flow through sand columns.

recognize that Eq. (3.1) leads to a multidimensional differential equation. Slichter stated that, in the absence of external forces such as gravity, the fluid velocity in a porous medium is proportional to the pressure gradient. He asserted that the constant of proportionality "depends upon the size of the soil grains, the porosity of the soil, and the viscosity of the liquid." He also added a term to accommodate the effects of gravity, combined the differential form of Darcy's law with the mass balance equation to obtain the single-phase flow equation, and developed solutions to the flow equation valid near wells.

Although some authors treat Darcy's law as a purely phenomenological observation, one can derive Darcy's law for a single fluid from the multiconstituent momentum balance reviewed in Section 2.5. The derivation given here comes with three caveats. First, it is one of many—following a wide variety of intellectual traditions—that have appeared in the scientific literature during the past half century. For a concise summary of these traditions, see [18]. Second, the presentation below glosses over several technical points for which details can be found in references cited throughout this section. Third, for simplicity's sake, we adopt assumptions that are not strictly necessary to derive Darcy's law.

3.1.1 Fluid Momentum Balance

The derivation begins with momentum balance equations for the fluid F and rock R, regarded as constituents in a multiphase mixture:

$$\phi_F \gamma_F \frac{D^F \mathbf{v}_F}{Dt} - \nabla \cdot \mathsf{T}_F - \phi_F \gamma_F \mathbf{b}_F = \mathbf{m}_F - \mathbf{v}_F \, r_F,$$

$$\phi_R \gamma_R \frac{D^R \mathbf{v}_R}{Dt} - \nabla \cdot \mathsf{T}_R - \phi_R \gamma_R \mathbf{b}_R = \mathbf{m}_R - \mathbf{v}_R \, r_R.$$

Three assumptions simplify the system:

Figure 3.2 A piece of mathematical surface in a porous medium, in which shaded areas represent the solid phase. Only a fraction of the surface area—the unshaded area—is available for fluid–fluid traction.

1. The rock is chemically inert, so $r_R = 0$. It follows from the total mass balance (2.30) that $r_F = 0$ as well.
2. The rock forms a rigid matrix. In this case, there exists a coordinate system in which $\mathbf{v}_R = \mathbf{0}$, and there is no need to solve the momentum balance equation for the rock.
3. The fluid flow is slow, in the sense that fluid acceleration is negligible compared with other terms in the fluid momentum balance. This assumption allows us to neglect the inertial term $D^F \mathbf{v}_F / Dt$ for the fluid phase.

One can derive Darcy's law without imposing the first two assumptions. Later in this chapter, we briefly discuss a model that relaxes the third assumption.

These three modeling assumptions leave us with a fluid momentum balance having the following form:

$$-\nabla \cdot \mathsf{T}_F - \phi_F \gamma_F \mathbf{b}_F = \mathbf{m}_F. \tag{3.2}$$

The rest of this chapter uses streamlined notation, dropping the subscript F and calling ϕ the **porosity** of the rock.

3.1.2 Constitutive Laws for the Fluid

Further headway requires constitutive relationships for the fluid stress tensor T, the momentum exchange rate $\mathbf{m}$, and the body force $\mathbf{b}$.

For the fluid stress tensor, we assume that the fluid is Newtonian:

$$\mathsf{T} = -\phi\, p\, \mathsf{I} + \phi \mu [\nabla \mathbf{v} + (\nabla \mathbf{v})^\top]. \tag{3.3}$$

Equation (3.3) differs from Eq. (2.14) by the factor ϕ. To justify this factor, consider a piece of smooth mathematical surface in the multiconstituent continuum, as drawn in Figure 3.2. On such a surface, fluid–fluid traction occurs only on that portion of the surface occupied by fluid, which we assume to be ϕ [48, p. 7].

The viscous stress term in Eq. (3.3) also deserves attention. Viscosity enables momentum transfer from regions where the velocity is large in magnitude to regions where it is smaller. In other words, momentum transfer moves down the velocity gradient. Figure 3.3 depicts this idea for a solid moving through a viscous fluid, in which the solid's motion imparts momentum to fluid at spatial

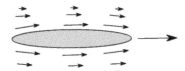

Figure 3.3 The transfer of momentum away from a solid object moving through a viscous fluid.

points distant from the solid boundary. We assume that the rate at which this type of momentum transfer occurs within the fluid is negligible compared with momentum losses to friction between the fluid and the rock. This assumption leaves us with the constitutive relationship

$$\mathsf{T} = -\phi \, p \, \mathsf{I}.$$

With this model for the fluid stress tensor T, the fluid momentum equation (3.2) becomes

$$-\nabla \cdot \mathsf{T} = \nabla \cdot (\phi p \mathsf{I}) = \underbrace{\phi \nabla p}_{(\mathrm{I})} + \underbrace{p \nabla \phi}_{(\mathrm{II})} = \phi \gamma \mathbf{b} + \mathbf{m}. \tag{3.4}$$

The term labeled (I) in Eq. (3.4) implies fluid flow in response to nonzero pressure gradients, a phenomenon that we expect to see in nature. However, the term labeled (II) appears to imply flow in response solely to nonzero porosity gradients, a phenomenon that we neither expect nor observe in flows through porous media. We discuss the resolution of this apparent paradox shortly.

For the momentum exchange, we adopt the constitutive relationship

$$\mathbf{m} = \underbrace{-\Lambda^{-1}\mathbf{v}}_{(\mathrm{III})} + \underbrace{p\nabla\phi}_{(\mathrm{IV})}. \tag{3.5}$$

The term labeled (III) resembles the Stokes drag on a solid sphere, discussed in Section 2.3. The nonnegative factor Λ, called the **resistivity**, quantifies the loss of momentum from the fluid attributable to friction exerted by solid grains in the porous medium. By analogy with the expression (2.25) for Stokes drag, we expect Λ to be proportional to the reciprocal $1/\mu$ of the fluid viscosity and to depend on the geometry of the rock at the microscopic scale of observation. Section 3.1.3 explores this intuition.

Justifying the term labeled (IV) in the constitutive relationship (3.5) requires more subtle reasoning. Fortuitously, this term exactly cancels the term (II) that arises from the Newtonian model (3.3) for the fluid stress tensor, thereby eliminating the apparent paradox of flow driven by porosity gradients. This welcome observation notwithstanding, a rigorous explanation of the origin of term (IV) requires a deeper thermodynamic analysis of the constitutive relationships than we undertake here. Iranian-born engineer Majid Hassanizadeh and American engineer William Gray were among the first to publish such an analysis [69], showing that a continuum version of the second law of thermodynamics prohibits flow driven by porosity gradients alone. For details, see [27, 69], [4, Chapter 7].

For the body force, we adopt a porosity-modified version of the expression given in Section 2.3:

$$\phi\gamma\mathbf{b} = \phi\gamma g\nabla Z,$$

where $Z(\mathbf{x})$ stands for depth below some datum and g is the gravitational acceleration. The factor ϕ in this relationship ensures that the body force on the fluid tends to zero as $\phi \to 0$ and to $\gamma g\nabla Z$ as $\phi \to 1$.

Substituting all of these constitutive relationships into Eq. (3.2) yields

$$\phi\nabla p - \phi\gamma g\nabla Z = -\Lambda^{-1}\mathbf{v},$$

or

$$\mathbf{v} = -\Lambda\phi(\nabla p - \gamma g\nabla Z). \tag{3.6}$$

This is **Darcy's law**.

It is useful to reflect on the significant simplifying assumptions required to derive Darcy's law from the momentum balance. Prominent among them are the following:

- The solid matrix is rigid.
- The inertial terms are negligible.
- Momentum transfer via shear stress in the fluid is negligible.
- The fluid loses momentum to the rock via Stokes drag, a phenomenon modeled in Section 2.4 as a time-independent effect.

These assumptions limit the applicability of Darcy's law to nearly steady, non-turbulent flows through porous media. Although this regime prevails in many underground flow applications, it does not adequately describe all flows through porous media. Section 3.2 reviews common extensions of Darcy's law that relax some of these assumptions.

3.1.3 Filtration Velocity

Groundwater hydrologists and petroleum engineers commonly use the quantity $\phi\mathbf{v}$, sometimes called the **filtration velocity**, **Darcy velocity** or **specific discharge**, instead of $\mathbf{v}$, which Todd [148, p. 67] calls the **average interstitial velocity**. The filtration velocity is the velocity-like quantity obtained by dividing the volumetric rate of fluid discharge, having dimension L^3T^{-1}, across a surface by the area of the surface, having dimension L^2. This quantity, easy to measure on the laboratory bench at macroscopic scales, is slower than the average insterstitial velocity **v**. The difference arises because only a fraction ϕ of the volume is accessible to the fluid, so the volumetric discharge rate is smaller than it would be were the entire cross section available for fluid flow. In terms of the filtration velocity, Darcy's law (3.6) becomes

$$\phi \mathbf{v} = -\Lambda \phi^2 (\nabla p - \gamma g \nabla Z). \tag{3.7}$$

3.1.4 Permeability

As mentioned, the reasoning about Stokes drag presented in Section 2.3 suggests that the coefficient Λ in Darcy's law (3.7) incorporates effects associated not only with the microscopic rock geometry but also effects associated with the fluid's viscosity. Consistent with this observation, in papers discussing the measurement of flow properties in porous media, American engineers P.G. Nutting [110] and R.D. Wyckoff et al. [160] proposed the following reformulation of Darcy's law:

$$\boxed{\phi \mathbf{v} = -\frac{k}{\mu} (\nabla p - \gamma g \nabla Z).} \tag{3.8}$$

The factor k, which can vary with spatial position, is the **permeability**.

Exercise 3.1 *Using Eq. (3.8), show that k has dimension* L^2.

Wyckoff and his coauthors proposed a unit of measurement for k that, in their view, would be practical for laboratory- and field-scale problems. Considering a one-dimensional, horizontal flow experiment with $\nabla Z = \mathbf{0}$, as drawn in Figure 3.4, they defined 1 **darcy** as the permeability required to allow fluid having viscosity $\mu = 1$ cP (water, for example) to discharge with filtration speed $\|\phi \mathbf{v}\| = 1$ cm s^{-1} in response to an applied pressure gradient $|dp/dx| = 1$ atm cm^{-1}, where 1 atm $= 101.325$ Pa is the average air pressure at sea level on Earth. For many natural sandstones, 10^{-3} darcies $\leqslant k \leqslant 1$ darcy. In SI units, 1 darcy is quite small, as the following exercise shows.

Exercise 3.2 *Show that*

$$1 \text{ darcy} = 0.987 \times 10^{-12} \text{ m}^2 \simeq 10^{-12} \text{ m}^2.$$

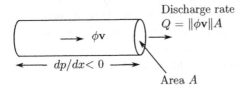

Discharge rate
$Q = \|\phi \mathbf{v}\| A$

Area A

Figure 3.4 Experimental configuration used to define 1 darcy.

Figure 3.5 Schematic diagram of an array of fixed solid in which viscous effects play a significant role in momentum transfer within the fluid, motivating the Brinkman law.

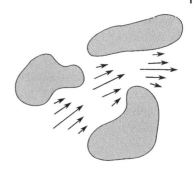

3.2 Non-Darcy Flows

As the derivation in Section 3.1 makes clear, Darcy's law rests on several assumptions about the nature of the fluid flow in the microscopic interstices of the rock matrix. We suspect that Darcy's law may fail, for example, when any of the following conditions holds.

1. The pores are so large that momentum transfer within the fluid phase is significant, as illustrated in Figure 3.5.
2. Inertial effects within the fluid are significant.
3. Fluid flows within the microscopic pore channels fail to obey the no-slip boundary condition (2.19).

In all of these cases, **non-Darcy flow** may prevail.

This section briefly reviews three mathematical models of non-Darcy flow in porous media: the Brinkman law, which includes the effects of viscosity on momentum transfer; the Forchheimer law, which incorporates inertial effects; and the Klinkenberg effect, which allows fluid to slip on the walls of the pores. Non-Darcy flows also occur in multiconstituent continua, such as fluid-saturated shales having pore diameters in the nanometer range, in which electrochemical effects influence fluid movements. Mathematical models of fluid movements in these media remain an active field of inquiry; see, for example, [37, 81].

3.2.1 The Brinkman Law

In 1947, Dutch physicist Hendrik C. Brinkman [28] modified Darcy's law to accommodate condition 1 in a porous medium consisting of an array of fixed solid particles. If we keep the viscous terms in the constitutive relationship (3.3) for the fluid, the fluid stress tensor is

$$\mathsf{T} = -\phi\, p\, \mathsf{I} + \phi\mu[\nabla\mathbf{v} + (\nabla\mathbf{v})^{\mathsf{T}}].$$

$$(3.9)$$

Exercise 3.3 *Assume that the fluid motion is incompressible, so* $\nabla \cdot \mathbf{v} = 0$, *and that the viscosity and porosity are uniform. Show that the constitutive relationship* (3.9) *yields*

$$\nabla \cdot \mathsf{T} = -\nabla(\phi p) + \phi \mu \nabla^2 \mathbf{v}.$$

We retain the other constitutive assumptions used to derive Darcy's law, namely the Stokes drag model (3.5),

$$\mathbf{m} = -\Lambda^{-1}\mathbf{v} + p\nabla\phi = -\frac{\phi^2 \mu}{k}\mathbf{v} + p\nabla\phi,$$

and the gravitational body force,

$$\mathbf{b} = g\nabla Z. \tag{3.10}$$

Substitution into the momentum balance (3.2) for the fluid phase yields the **Brinkman law**,

$$-\nabla p + \mu \nabla^2 \mathbf{v} + \gamma g \nabla Z = \frac{\mu}{k}\phi \mathbf{v}.$$

Theoretical evidence [9] suggests that the Brinkman law has limited validity in geologic porous media *per se* but may be applicable in swarms of unconnected particles, such as proppants used in hydraulic fracturing [80], and in some fibrous porous media.

3.2.2 The Forchheimer Equation

To accommodate condition 2, Philipp Forchheimer [54] proposed an approach that does not explicitly include the inertial terms in the momentum balance. Instead, it allows for higher order contributions of the fluid velocity to the Stokes drag model. Specifically, Forchheimer modeled the one-dimensional fluid response to a pressure drop along a horizontal porous medium, as drawn in Figure 3.6, using an equation of the form

$$\frac{p_2 - p_1}{x_2 - x_1} = \frac{\mu}{k}\phi v + \beta \gamma (\phi v)^2,$$

where $\beta > 0$. To extend this model to three dimensions, adopt a momentum exchange term of the following form:

$$\mathbf{m} = F(\|\phi \mathbf{v}\|)\phi \mathbf{v} + p\nabla\phi,$$

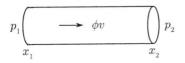

Figure 3.6 One-dimensional flow geometry used to motivate the Forchheimer equation.

representing the function F in a power-series expansion about $\|\phi\mathbf{v}\| = 0$:

$$F(\|\phi\mathbf{v}\|) = \frac{\mu}{k} + \beta\gamma\|\phi\mathbf{v}\| + \mathcal{O}(\|\phi\mathbf{v}\|^2).$$

The notation $f(y) = \mathcal{O}(y)$ means that there exists a positive constant M such that $|f(y)| \leqslant M|y|$ whenever $|y|$ is sufficiently small.

Neglecting the inertial terms in the momentum balance (3.2), retaining only the terms through the first power in $\|\phi\mathbf{v}\|$ in F, and adopting Eq. (3.10) for the body force yields the **Forchheimer equation**,

$$\left(\frac{\mu}{k} + \beta\gamma\|\phi\mathbf{v}\| \right) \phi\mathbf{v} = -(\nabla p - \gamma g \nabla Z). \tag{3.11}$$

The parameter $1/\beta$ is sometimes called the **inertial permeability**. Equation (3.11) is inherently nonlinear in the fluid velocity $\mathbf{v}$.

A common engineering criterion for determining whether condition 2 applies is that the **pore Reynolds number** $\mathrm{Re} = \gamma\|\mathbf{v}\|d/\mu > 40$, where d denotes the mean grain diameter of the rock [163].

3.2.3 The Klinkenberg Effect

The flow of a low-viscosity gas such as air or nitrogen in geologic porous media exhibits an effect that one must take into account when modeling low-pressure, low-viscosity flows in porous media. In 1941, Dutch chemist L.J. Klinkenberg [90] noted that permeabilities measured using low-pressure gases are larger in magnitude than those measured using more viscous fluids, such as brine. He attributed this effect to a failure of the no-slip boundary condition (2.19) when gases flow through the channels of a geologic porous medium. He proposed modeling the effect of gas slipping along the solid walls of the pores using the following relationship between the measured gas permeability k_G and the permeability k used in Eq. (3.8):

$$k_G = k \left(1 + \frac{b}{p} \right). \tag{3.12}$$

Here, p stands for the pressure, and b denotes a nonnegative constant determined empirically for the gas in question.

This **Klinkenberg effect** plays a significant role in flows in which the mean free path of the fluid molecules is comparable in length to the average pore diameter, as one expects in low-density flows through fine-grained porous media. The effect may be important in the interpretation of core analyses in unconventional oil and gas reservoirs involving low-permeability rocks. For a compendium of solutions to flow equations involving Eq. (3.12), see [158].

Although these non-Darcy effects, modeled by the Brinkman law, the Forchheimer equation, and the Klinkenberg effect, play significant roles in some settings, the remainder of this book focuses on porous media in which Darcy's law and its multifluid extensions furnish accurate models.

3.3 The Single-fluid Flow Equation

With Darcy's field equation (3.8) for single-fluid flow in hand, we now develop a flow equation. The derivation follows a pattern seen frequently in fluid mechanics: Substitute Darcy's law into the mass balance

$$\frac{\partial}{\partial t}(\phi\gamma) + \nabla \cdot (\phi\gamma\mathbf{v}) = 0 \tag{3.13}$$

to obtain a second-order partial differential equation (PDE). This straightforward tactic yields

$$\frac{\partial}{\partial t}(\phi\gamma) - \nabla \cdot \left[\frac{\gamma k}{\mu}(\nabla p - \gamma g \nabla Z)\right] = 0.$$

However, in keeping with traditional groundwater hydrology, the development presented in this section involves more than a straightforward substitution, allowing for small deformations of the rock matrix as well as effects arising from fluid compressibility. This approach relaxes the assumption in Section 3.1 that the rock matrix is rigid.

Exercise 3.4 *Show that the mass balance equation* (3.13) *for the fluid has the equivalent form*

$$\phi\frac{D^F\gamma}{Dt} + \gamma\frac{\partial\phi}{\partial t} + \gamma\nabla \cdot (\phi\mathbf{v}) = 0. \tag{3.14}$$

Similarly, we write the mass balance for the rock in the form

$$\phi\frac{D^R\gamma_R}{Dt} + \gamma_R\frac{\partial}{\partial t}(1 - \phi) + \gamma_R\nabla \cdot [(1 - \phi)\mathbf{v}_R] = 0.$$

If the solid grains are incompressible—which is different from assuming that the solid matrix composed of the grains is incompressible—then $D^R\gamma_R/Dt = 0$. This assumption, the identity

$$\frac{\partial(1 - \phi)}{\partial t} = -\frac{\partial\phi}{\partial t},$$

and the fact that $\gamma_R \neq 0$ yield

$$\nabla \cdot [(1 - \phi)\mathbf{v}_R] = \frac{\partial\phi}{\partial t}. \tag{3.15}$$

We now add the two mass balance equations (3.14) and (3.15). Decomposing $\mathbf{v} = \mathbf{v}_R + \mathbf{v}_\Delta$ allows us to rewrite Eq. (3.14) as follows:

$$\phi \frac{D^F \gamma}{Dt} + \underbrace{\gamma \frac{\partial \phi}{\partial t}}_{(I)} + \underbrace{\gamma \nabla \cdot [\phi(\mathbf{v}_R + \mathbf{v}_\Delta)]}_{(II)} = 0.$$

By Eq. (3.15), the term labeled (I) is $\gamma \nabla \cdot \mathbf{v}_R - \gamma \nabla \cdot (\phi \mathbf{v}_R)$. The term labeled (II) is $\gamma \nabla \cdot (\phi \mathbf{v}_R) + \gamma \nabla \cdot (\phi \mathbf{v}_\Delta)$. Adding and taking advantage of cancellation yields the combined mass balance equation,

$$\phi \frac{D^F \gamma}{Dt} + \gamma \nabla \cdot \mathbf{v}_R + \gamma \nabla \cdot (\phi \mathbf{v}_\Delta) = 0. \tag{3.16}$$

Exercise 3.5 *Show that* $(1 - \phi)\gamma \nabla \cdot \mathbf{v}_R = \gamma D^R \phi / Dt$. *Use this identity and Eq.* (3.16) *to obtain the equation*

$$\phi \frac{D^F \gamma}{Dt} + \frac{\gamma}{1 - \phi} \frac{D^R \phi}{Dt} + \gamma \nabla \cdot (\phi \mathbf{v}_\Delta) = 0. \tag{3.17}$$

3.3.1 Fluid Compressibility and Storage

Further headway requires additional constitutive assumptions about the fluid and rock. For the fluid, we assume a differentiable **equation of state** $\gamma = \gamma(p)$, where $\gamma'(p) > 0$, so density increases with pressure. By the chain rule,

$$\frac{D^F \gamma}{Dt} = \gamma'(p) \frac{D^F p}{Dt} = \beta \gamma \frac{D^F p}{Dt}, \tag{3.18}$$

$$\nabla \gamma = \beta \gamma \nabla p. \tag{3.19}$$

Here, the coefficient $\beta = \gamma' / \gamma$, a positive function of pressure, denotes the **fluid compressibility**, having dimension $M^{-1}LT^2$. For the rock, we allow for compressibility of the matrix (not the grains) by assuming that $\phi = \phi(p)$, with $\phi'(p) \geq 0$. Another application of the chain rule yields

$$\frac{D^R \phi}{Dt} = \phi'(p) \frac{D^R p}{Dt} = (1 - \phi)\alpha \frac{D^R p}{Dt}, \tag{3.20}$$

where the coefficient $\alpha = \phi' / [1 - \phi]$ also has dimension $M^{-1}LT^2$.

Substituting the expressions (3.18)–(3.20) into the combined mass balance (3.17) and dividing through by γ, we get

$$\phi \beta \frac{D^F p}{Dt} + \alpha \frac{D^R p}{Dt} + \nabla \cdot (\phi \mathbf{v}_\Delta) = 0. \tag{3.21}$$

In many groundwater flows, the rock and fluid velocities are small, as are pressure gradients. In these flows, the velocity-driven components of the two material derivatives appearing in Eq. (3.21) are products of small quantities and hence negligible: $\mathbf{v}_R \cdot \nabla p \simeq 0 \simeq \mathbf{v} \cdot \nabla p$. Neglecting these terms yields the final form for the combined fluid-rock mass balance equation,

$$S_p \frac{\partial p}{\partial t} + \nabla \cdot (\phi \mathbf{v}_\Delta) = 0. \tag{3.22}$$

Here, the factor $S_p = \phi\beta + \alpha$, a positive function of pressure, captures the combined effect of fluid and rock-matrix compressibility, an effect that hydrologists call **storage**.

3.3.2 Combining Darcy's Law and the Mass Balance

At last, we substitute Darcy's law, in the form

$$\phi \mathbf{v}_\Delta = -\frac{k}{\mu}(\nabla p - \gamma g \nabla Z)$$

into the mass balance (3.22), obtaining

$$\boxed{S_p \frac{\partial p}{\partial t} - \nabla \cdot \left[\frac{k}{\mu}(\nabla p - \gamma g \nabla Z) \right] = 0.} \tag{3.23}$$

The second-order PDE (3.23) is the **single-fluid flow equation**, written in terms of fluid pressure p.

3.4 Potential Form of the Flow Equation

In two classic papers, American geophysicist M. King Hubbert [75, 76] examined conditions under which one can express the filtration velocity $\phi\mathbf{v}$ as the gradient of a scalar-valued function of position. Hubbert began by writing Darcy's law as

$$\phi \mathbf{v} = -\frac{k}{\mu}(\nabla p - \gamma g \nabla Z) = \frac{\gamma k}{\mu}\mathbf{E}, \tag{3.24}$$

where

$$\mathbf{E} = g\nabla Z - \frac{1}{\gamma}\nabla p$$

signifies the force per unit mass (LT^{-2}) acting on the fluid. Using this notation, we seek conditions under which

$$\mathbf{E} = -\nabla\Phi, \tag{3.25}$$

for some **scalar potential** Φ, which must have dimension $L^2 T^{-2}$. For notational convenience, in this section, we temporarily suppress the dependence on time.

3.4.1 Conditions for the Existence of a Potential

A necessary condition for a relationship of the form (3.25) to hold for a differentiable vector field $\mathbf{E}$ is that the curl of the vector field $\mathbf{E}$ vanish (see [99, Section 8.3]). In Cartesian coordinates, this condition requires that

$$\nabla \times \mathbf{E} = \left(\frac{\partial E_3}{\partial x_2} - \frac{\partial E_2}{\partial x_3}, \frac{\partial E_1}{\partial x_3} - \frac{\partial E_3}{\partial x_1}, \frac{\partial E_2}{\partial x_1} - \frac{\partial E_1}{\partial x_2} \right) = \mathbf{0}.$$

Exercise 3.6 *Review conditions under which $\nabla \times \mathbf{E} = \mathbf{0}$ is a sufficient condition for $\mathbf{E}$ to be a gradient.*

If the gravitational acceleration g is constant, then, by the product rule,

$$\nabla \times \mathbf{E} = g\nabla \times \nabla Z - \nabla \times \left(\frac{1}{\gamma} \nabla p \right)$$

$$= -\nabla \left(\frac{1}{\gamma} \right) \times \nabla p - \frac{1}{\gamma} \nabla \times \nabla p$$

$$= -\nabla \left(\frac{1}{\gamma} \right) \times \nabla p,$$

since the curl of a differentiable gradient vanishes. This identity reveals three conditions under which $\nabla \times \mathbf{E} = \mathbf{0}$:

1. $\nabla p = \mathbf{0}$, a case of little interest.
2. The fluid density γ is constant.
3. The vector field $\nabla(1/\gamma)$ is parallel to the vector field ∇p.

In the last two cases, the fact that the gradient of a function is orthogonal to its level sets, as illustrated in Figure 3.7, implies that level sets of p are also level sets of γ. In other words, $\mathbf{E} = -\nabla\Phi$ only if $\gamma = \gamma(p)$, including the case in which γ is constant. We call a fluid whose density depends only on pressure a **barotropic fluid**. The hypothesis that the fluid is barotropic excludes nonisothermal flows, in which γ is a function of pressure and temperature, and compositional flows, for which the fluid density may also depend on chemical composition. Chapter 7 introduces models of compositional flows.

Figure 3.7 Level sets of $p(\mathbf{x})$, shown as dashed curves, along with the gradient $\nabla p(\mathbf{x})$ at a point, showing that the gradient is orthogonal to level sets.

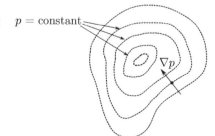

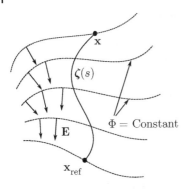

Figure 3.8 Construction of a scalar potential $\Phi(\mathbf{x})$ such that $\mathbf{E} = -\nabla\Phi$, by computing a path integral along ζ from a reference point $\mathbf{x}_{ref}$ to $\mathbf{x}$.

3.4.2 Calculating the Scalar Potential

Next we examine how to calculate the scalar potential Φ in cases where it exists. Because only its gradient is of interest, we expect to define Φ up to an additive constant.

Pick a reference spatial position $\mathbf{x}_{ref}$, at which we can assign the depth $Z(\mathbf{x}_{ref}) = -\mathbf{x}_{ref} \cdot \mathbf{e}_3 = 0$. To define $\Phi(\mathbf{x})$ at an arbitrary spatial position $\mathbf{x}$, let ζ be any continuously differentiable path in the porous medium such that $\zeta(0) = \mathbf{x}_{ref}$, $\zeta(1) = \mathbf{x}$, and $\zeta'(s) \neq \mathbf{0}$ for all $s \in [0, 1]$, as drawn in Figure 3.8.

By the fundamental theorem of calculus and the definition of $\mathbf{E}$,

$$
\begin{aligned}
\Phi(\mathbf{x}) &= \Phi(\mathbf{x}_{ref}) + \int_\zeta \nabla\Phi \cdot d\mathbf{x} \\
&= \Phi(\mathbf{x}_{ref}) - \int_0^1 \mathbf{E}(\zeta(s)) \cdot \zeta'(s)\, ds \\
&= \Phi(\mathbf{x}_{ref}) \underbrace{- g \int_0^1 \nabla Z(\zeta(s)) \cdot \zeta'(s)\, ds}_{(I)} + \underbrace{\int_0^1 \frac{\nabla p(\zeta(s))}{\gamma(p(\zeta(s)))} \cdot \zeta'(s)\, ds}_{(II)}.
\end{aligned}
\tag{3.26}
$$

The terms labeled (I) and (II) in this equation simplify as follows: By the chain rule,

$$
(I) = -g \int_0^1 \frac{d}{ds} Z(\zeta(s))\, ds = -g[Z(\mathbf{x}) - Z(\mathbf{x}_{ref})] = -gZ(\mathbf{x}),
$$

and by the change of variables $u = p(\zeta(s))$,

$$
(II) = \int_0^1 \frac{1}{\gamma(p(\zeta(s)))} \frac{dp}{ds}(\zeta(s))\, ds = \int_{p(\mathbf{x}_{ref})}^{p(\mathbf{x})} \frac{du}{\gamma(u)}.
\tag{3.27}
$$

The value of the last integral in Eq. (3.27) depends only on the endpoint values $p(\mathbf{x}_{ref})$ and $p(\mathbf{x})$ and is therefore independent of the path taken from $\mathbf{x}_{ref}$ to $\mathbf{x}$. Also, we can select any value we wish for $\Phi(\mathbf{x}_{ref})$, say $\Phi(\mathbf{x}_{ref}) = 0$, without affecting $\nabla\Phi$.

Substituting these values into Eq. (3.26) yields the **Hubbert potential**,

$$\Phi(\mathbf{x}) = -gZ(\mathbf{x}) + \int_{p(\mathbf{x}_{\text{ref}})}^{p(\mathbf{x})} \frac{du}{\gamma(u)}, \tag{3.28}$$

having dimension energy/mass, or $L^2 T^{-2}$.

The Hubbert potential $\Phi(\mathbf{x})$ gives the mechanical energy per unit mass required to move the fluid from the reference position $\mathbf{x}_{\text{ref}}$ to position $\mathbf{x}$. In the special case where the fluid density γ is constant, which is approximately true in many groundwater aquifers, Eq. (3.28) simplifies to

$$\Phi = -gZ + \frac{p - p_{\text{ref}}}{\gamma},$$

where $p_{\text{ref}} = p(\mathbf{x}_{\text{ref}})$ is the pressure at the reference position.

In summary, for barotropic fluids,

$$\mathbf{E} = \frac{\mu}{\gamma k} \phi \mathbf{v} = -\nabla\Phi.$$

Solving for the filtration velocity yields the following form of Darcy's law:

$$\phi \mathbf{v} = -\frac{\gamma k}{\mu} \nabla\Phi.$$

3.4.3 Piezometric Head

Groundwater hydrologists seldom use the Hubbert potential, preferring a closely related scalar potential having the form

$$\boxed{H(\mathbf{x}) = \frac{\Phi(\mathbf{x})}{g} = -Z(\mathbf{x}) + \frac{1}{g} \int_{p(\mathbf{x}_{\text{ref}})}^{p(\mathbf{x})} \frac{du}{\gamma(u)}.} \tag{3.29}$$

The function $H(\mathbf{x})$, having dimension L, is the **piezometric head**. It enjoys a useful practical interpretation: $H(\mathbf{x})$ gives the height above some datum $Z = 0$ to which fluid rises in a **piezometer** tapped into the fluid-saturated porous medium at the point $\mathbf{x}$, as depicted in Figure 3.9. When the fluid density γ is constant,

$$H = -Z + \frac{p - p_{\text{ref}}}{\gamma g},$$

where p_{ref} is the pressure at the datum.

Exercise 3.7 *Suppose that a water well is cased (that is, lined with an impermeable pipe) from Earth's surface down to 100 m above sea level, where a screen in the casing allows water to flow into the well from an aquifer. Water rises 10 m above the screen. If the datum elevation is sea level, what is the piezometric head at the well? What is Z?*

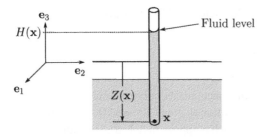

Figure 3.9 A piezometer showing the piezometric head $H(\mathbf{x})$ as the height to which fluid rises in a tube tapped into the porous medium at $\mathbf{x}$.

In terms of H, Darcy's law takes the simple form

$$\boxed{\phi\mathbf{v} = -K\nabla H,}$$
(3.30)

where $K = \gamma kg/\mu$ is the **hydraulic conductivity**, having dimension LT^{-1}.

3.4.4 Head-Based Flow Equation

We now recast the single-fluid flow equation (3.23) in terms of the piezometric head H. With the reformulation (3.30) of Darcy's law in hand, the main task is to reformulate the mass balance equation (3.22), written as

$$S_p\frac{\partial p}{\partial t} + \nabla\cdot(\phi\mathbf{v}) = 0.$$
(3.31)

We begin by deriving an expression for $\partial H/\partial t$, a step that requires differentiating the integral in Eq. (3.29). For simplicity's sake, assume that the fluid density γ is constant, which is approximately true for most groundwater aquifers, and that $p(\mathbf{x}_{\text{ref}})$ is constant, which is approximately true if we chose $\mathbf{x}_{\text{ref}}$ to be a point where the pressure is atmospheric.

Exercise 3.8 *Use the chain rule and the fundamental theorem of calculus to prove the **Leibniz rule**: For a function f of two variables,*

$$\frac{d}{d\xi}\int_{a(\xi)}^{b(\xi)} f(\xi,\eta)\,d\eta = \int_{a(\xi)}^{b(\xi)} \frac{\partial f}{\partial \xi}(\xi,\eta)\,d\eta$$
$$+ f(\xi,b(\xi))\frac{db}{d\xi}(\xi) - f(\xi,a(\xi))\frac{da}{d\xi}(\xi)$$
(3.32)

provided a, b, and f are differentiable.

Now use Eq. (3.32) to compute the required derivative:

$$\frac{\partial H}{\partial t} = -\underbrace{\frac{\partial Z}{\partial t}(\mathbf{x})}_{0} + \underbrace{\int_{p(\mathbf{x}_{\text{ref}})}^{p(\mathbf{x},t)} \frac{\partial}{\partial t}\left(\frac{1}{\gamma g}\right) du}_{0} + \frac{1}{\gamma g}\frac{\partial p}{\partial t}(\mathbf{x},t) - \underbrace{\frac{1}{\gamma g}\frac{\partial}{\partial t}(p(\mathbf{x}_{\text{ref}}))}_{0}.$$
(3.33)

The terms underscored with 0 vanish, being time derivatives of quantities that do not vary with time.

From Eq. (3.33), it follows that

$$\frac{\partial p}{\partial t} = \gamma g \frac{\partial H}{\partial t}.$$

This identity allows us to rewrite the mass balance equation (3.31) as follows:

$$S_s \frac{\partial H}{\partial t} + \nabla \cdot (\phi \mathbf{v}) = 0. \qquad (3.34)$$

Here, the **specific storage** $S_s = \gamma g S_p$, having dimension L^{-1}, is a positive function giving the volume of water released per unit volume of aquifer in response to a decrease in H by one unit. Substituting for the filtration velocity $\phi \mathbf{v}$ in Eq. (3.34) using Darcy's law in the form (3.30) yields the **head-based flow equation**,

$$\boxed{S_s \frac{\partial H}{\partial t} - \nabla \cdot (K \nabla H) = 0.} \qquad (3.35)$$

In steady flows, $\partial H/\partial t = 0$, and Eq. (3.35) reduces to the elliptic PDE

$$\nabla \cdot (K \nabla H) = 0. \qquad (3.36)$$

Exercise 3.9 *Show that the PDE (3.35) is parabolic at every point $\mathbf{x}$ where $K(\mathbf{x})$ is positive.*

3.4.5 Auxiliary Conditions for the Flow Equation

A problem involving a PDE is **well-posed** if it satisfies three conditions:

1. A solution to the problem exists.
2. The solution is unique.
3. The solution depends continuously on the data that define the problem.

The hypotheses needed to guarantee these three conditions vary with the type of PDE involved. Some PDEs admit no well-posed problems; see [65] for a thorough introduction. For those that do, it is generally necessary to prescribe **auxiliary conditions**, in the form of initial and boundary conditions, to guarantee uniqueness.

The time-dependent flow equation (3.35) generalizes the heat equation, one of the classic PDEs of mathematical physics. When the coefficients S_s and K are positive functions of position and bounded away from zero, the equation is parabolic, as shown in Exercise 3.9. For equations of this type on a bounded spatial domain Ω, well-posed problems require the following auxiliary conditions:

- An **initial condition**, that is, a prescribed function $H(\mathbf{x}, t_0)$ defined on Ω at some fixed initial time t_0.

- **Boundary conditions**, that is, known conditions satisfied by the solution on the boundary $\partial\Omega$.

Three types of boundary conditions commonly arise in connection with the flow equation. The most straightforward are **Dirichlet conditions**, which prescribe the values of $H(\mathbf{x}, t)$ on some subset $\Gamma_D \subset \partial\Omega$ for all times $t > t_0$:

$$H(\mathbf{x}, t) = H_\partial(\mathbf{x}, t), \quad \text{for } \mathbf{x} \in \Gamma_D \subset \partial\Omega, \ t > t_0. \tag{3.37}$$

Here, H_∂ is a known function.

Also useful in many problems are **Neumann conditions**, which prescribe the outward flux of fluid across some subset $\Gamma_N \subset \partial\Omega$. If $q(\mathbf{x})$ denotes the known component of the outward fluid flux orthogonal to the boundary at the point $\mathbf{x} \in \partial\Omega$, by Darcy's law in the form (3.30) Neumann conditions have the form

$$-K(\mathbf{x})\nabla H(\mathbf{x}, t) \cdot \mathbf{n}(\mathbf{x}) = q(\mathbf{x}), \quad \text{for } \mathbf{x} \in \Gamma_N \subset \partial\Omega, \ t > t_0. \tag{3.38}$$

Here, $\mathbf{n}(\mathbf{x})$ denotes the unit-length vector pointing outward from the spatial domain Ω and orthogonal to its bounding surface $\partial\Omega$ at $\mathbf{x}$. In the special case when no fluid flows across the boundary—for example, when $\partial\Omega$ is impermeable or when certain symmetry conditions hold—Eq. (3.38) reduces to the **no-flux boundary condition**,

$$-K(\mathbf{x})\nabla H(\mathbf{x}, t) \cdot \mathbf{n}(\mathbf{x}) = 0, \quad \text{for } \mathbf{x} \in \Gamma_N \subset \partial\Omega, \ t > t_0.$$

A third type of boundary condition, the **Robin condition**, occasionally appears. Robin conditions model boundaries across which fluid can leak in response to differences in piezometric head between the interior and exterior of Ω, for example when the confining rock is semipermeable:

$$-K(\mathbf{x})\nabla H(\mathbf{x}, t) \cdot \mathbf{n}(\mathbf{x}) = \kappa(\mathbf{x}) \left[H(\mathbf{x}, t) - H_{\text{ext}}(\mathbf{x}, t) \right],$$
$$\text{for } \mathbf{x} \in \Gamma_R \subset \partial\Omega, \ t > t_0.$$

Here, κ denotes a known, nonnegative function characterizing the leakage, and H_{ext} is the exterior piezometric head, also assumed to be known. Rearranging gives

$$\kappa(\mathbf{x})H(\mathbf{x}, t) + K(\mathbf{x})\nabla H(\mathbf{x}, t) \cdot \mathbf{n}(\mathbf{x}) = \kappa(\mathbf{x})H_{\text{ext}}(\mathbf{x}, t),$$
$$\text{for } \mathbf{x} \in \Gamma_R \subset \partial\Omega, \ t > t_0. \tag{3.39}$$

The boundary segments in Eqs. (3.37)–(3.39) must completely account for the boundary: $\Gamma_D \cup \Gamma_N \cup \Gamma_R = \partial\Omega$.

In the special case of steady flows, the elliptic PDE (3.36) holds. No initial condition is required, but we must impose Dirichlet, Neumann, or Robin boundary conditions—or some combination of them—to ensure uniqueness of the solution. In this case, pure Neumann conditions, in which $\Gamma_N = \partial\Omega$, require care, for two reasons.

Exercise 3.10   *Suppose that H(**x**) is a solution to a boundary-value problem of the form*

$$\nabla \cdot (K\nabla H) = 0, \quad \text{for } \mathbf{x} \in \Omega;$$

$$-K\nabla H \cdot \mathbf{n} = q, \quad \text{for } \mathbf{x} \in \partial\Omega. \tag{3.40}$$

*Show that H(**x**) + C is also a solution for any constant C.*

Exercise 3.10 shows that the steady problem with pure Neumann boundary conditions does not have a unique solution. To ensure a well-posed problem, it is necessary to prescribe the value of H at at least one point on $\partial\Omega$.

Even if one can tolerate uniqueness only up to an additive constant, pure Neumann conditions must satisfy a constraint for the steady problem in Exercise 3.10 to possess any solutions.

Exercise 3.11   *Use the divergence theorem to derive the following condition on the prescribed net flux q for the boundary-value problem (3.40):*

$$\int_{\partial\Omega} q \, ds = 0.$$

The result of Exercise 3.11 shows that no steady solution can exist unless the net influx of fluid across $\partial\Omega$ balances the net outflux.

Wells constitute a special category of boundaries in many applications involving geologic porous media. Section 4.1 introduces this topic.

3.5   Areal Flow Equation

In some applications, the fluid flow is nearly two-dimensional. This approximation yields considerable computational benefits when the areal extent of the flow domain, for example a groundwater aquifer, greatly exceeds its thickness. Figure 3.10 depicts a **confined** aquifer for which low-permeability lower and upper confining layers bound the permeable rock formation between possibly time-dependent vertical coordinates $x_3 = a(x_1, x_2, t)$ and $x_3 = b(x_1, x_2, t)$, respectively, where $a(x_1, x_2, t) \neq b(x_1, x_2, t)$.

This section derives a two-dimensional flow equation that accommodates spatial and temporal variations in these two bounding surfaces. The derivation assumes that the material properties S_s, ϕ, and K do not vary as functions of x_3 and that the fluid velocity has no vertical component:

$$\frac{\partial S_s}{\partial x_3} = \frac{\partial \phi}{\partial x_3} = \frac{\partial K}{\partial x_3} = \mathbf{v} \cdot \mathbf{e}_3 = 0.$$

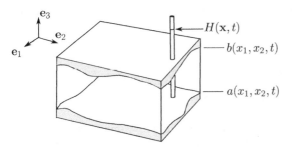

Figure 3.10 Geometry of areal flow in a confined aquifer. The elevations $a(x_1, x_2, t)$ and $b(x_1, x_2, t)$ of the lower and upper confining layers, respectively, may vary with areal position (x_1, x_2) and time t.

Define the **vertical average** of any integrable function $\varphi(\mathbf{x}, t)$ as follows:

$$\overline{\varphi}(x_1, x_2, t) = \frac{1}{l(x_1, x_2, t)} \int_a^b \varphi(x_1, x_2, x_3, t) \, dx_3,$$

where $l(x_1, x_2, t) = b(x_1, x_2, t) - a(x_1, x_2, t)$ denotes the nonzero thickness of the formation. We apply the vertical averaging operator to the terms in the three-dimensional mass balance,

$$S_s \frac{\partial H}{\partial t} + \nabla \cdot (\phi \mathbf{v}) = 0,$$

and Darcy's law,

$$\phi \mathbf{v} + K \nabla H = 0, \tag{3.41}$$

then combine the results to derive an areal flow equation.

3.5.1 Vertically Averaged Mass Balance

Averaging the mass balance and multiplying by l yields

$$\int_a^b \left[S_s \frac{\partial H}{\partial t} + \nabla \cdot (\phi \mathbf{v}) \right] dx_3 = 0.$$

Since $\partial S_s / \partial x_3 = 0$, applying the Leibniz rule (3.32) to the integral yields

$$\int_a^b S_s \frac{\partial H}{\partial t} \, dx_3 = S_s \int_a^b \frac{\partial H}{\partial t} \, dx_3$$

$$= S_s \left[\frac{\partial}{\partial t} \int_a^b H \, dx_3 - H \Big|_{x_3=b} \frac{\partial b}{\partial t} + H \Big|_{x_3=a} \frac{\partial a}{\partial t} \right]$$

$$= S_s \left[\frac{\partial}{\partial t} (l \overline{H}) - H \Big|_{x_3=b} \frac{\partial b}{\partial t} + H \Big|_{x_3=a} \frac{\partial a}{\partial t} \right]. \tag{3.42}$$

In the following exercise and for most of this section, we temporarily use the symbol ∇_2, having vector representation

$$\left(\frac{\partial}{\partial x_1}, \frac{\partial}{\partial x_2} \right)$$

to distinguish the two-dimensional gradient operator from the three-dimensional gradient operator. Afterward, we revert to the symbol ∇ and rely on context to distinguish the two- and three-dimensional cases.

Exercise 3.12 *Show that*

$$\int_a^b \nabla \cdot (\phi \mathbf{v}) \, dx_3 = \nabla_2 \cdot \int_a^b \phi \mathbf{v} \, dx_3$$
$$+ \phi \mathbf{v}\Big|_{x_3=b} \cdot \nabla(x_3 - b) - \phi \mathbf{v}\Big|_{x_3=a} \cdot \nabla(x_3 - a)$$
$$= \nabla_2 \cdot (\overline{l\phi \mathbf{v}}) + \phi \mathbf{v}\Big|_{x_3=b} \cdot \nabla(x_3 - b) - \phi \mathbf{v}\Big|_{x_3=a} \cdot \nabla(x_3 - a).$$

Equation (3.42) and the result of Exercise 3.12 yield the following form for the vertically averaged mass balance:

$$S_s \left[\frac{\partial}{\partial t}(l\overline{H}) - H\Big|_{x_3=b} \frac{\partial b}{\partial t} + H\Big|_{x_3=a} \frac{\partial a}{\partial t} \right] + \nabla_2 \cdot (\overline{l\phi \mathbf{v}}) + q_t - q_b, \tag{3.43}$$

where

$$q_t = \phi \mathbf{v} \cdot \nabla(x_3 - b),$$
$$q_b = \phi \mathbf{v} \cdot \nabla(x_3 - a). \tag{3.44}$$

As Figure 3.11 illustrates, the vector fields $\nabla(x_3 - a)$ and $\nabla(x_3 - b)$ are orthogonal to the level sets $x_3 - a = 0$ and $x_3 - b = 0$, respectively. By this reasoning the expressions in Eqs. (3.44) represent the net fluid flux perpendicular to the low-permeability surfaces bounding the top and bottom, respectively, of the aquifer. Hydrologists refer to q_t and q_b as **leakage** terms.

It is common to adopt the additional modeling assumption that vertical variations in piezometric head are negligible:

$$H\Big|_{x_3=b} \simeq H\Big|_{x_3=a} \simeq \overline{H}. \tag{3.45}$$

This assumption restricts the validity of the analysis to thin aquifers in which vertical variations in piezometric head across the thickness of the aquifer are much less significant than variations attributable to changes in the depth of the aquifer

Figure 3.11 Fluid flux across the upper confining layer.

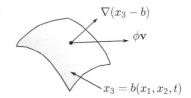

and to variations imposed by pumping at wells. In this case, since $l = b - a$, the vertically averaged mass balance (3.43) reduces to

$$S_s \left[\frac{\partial}{\partial t}(l\overline{H}) - \overline{H}\frac{\partial l}{\partial t} \right] + \nabla_2 \cdot (l\overline{\phi v}) + q_t - q_b = 0,$$

or, by the product rule,

$$S\frac{\partial \overline{H}}{\partial t} + \nabla_2 \cdot (l\overline{\phi v}) + q_t - q_b = 0. \tag{3.46}$$

Here, $S = lS_s$ is a dimensionless function called the **storativity**, representing the volume of fluid released per unit area of the aquifer as a result of a unit decrease in piezometric head.

3.5.2 Vertically Averaged Darcy's Law

Vertically integrating Darcy's law (3.41) yields

$$\int_a^b (\phi v + K\nabla H) \, dx_3 = 0. \tag{3.47}$$

The first term on the left side of Eq. (3.47) reduces as follows:

$$\int_a^b \phi v \, dx_3 = l\overline{\phi v}. \tag{3.48}$$

To calculate the integral of the second term on the left side of Eq. (3.47), use the assumption that $\partial K/\partial x_3 = 0$, combined with the assumption (3.45) about vertical variations in piezometric head:

$$\int_a^b K\nabla H \, dx_3 = K \left(\nabla_2 \int_a^b H \, dx_3 - \overline{H}\nabla_2 b + \overline{H}\nabla_2 a \right)$$

$$= K \left[\nabla_2 (l\overline{H}) - \overline{H}\nabla_2 l \right] = lK\nabla_2\overline{H},$$

since $l = b - a$. Combining this equation with the result (3.48) gives the following form for the vertically averaged version of Darcy's law:

$$l\overline{\phi v} + T\nabla_2\overline{H} = 0, \tag{3.49}$$

where the coefficient $T = lK$, having dimension L^2T^{-1}, is the **transmissivity**.

Finally, we substitute Darcy's law in the form (3.49) into the vertically averaged mass balance (3.46) to arrive at the **areal flow equation**:

$$\boxed{S\frac{\partial H}{\partial t} - \nabla \cdot (T\nabla H) = q_b - q_t.} \tag{3.50}$$

(Now that there is no risk of ambiguity, we drop the subscript from the two-dimensional gradient operator and the overbar notation for vertical average.)

This parabolic PDE has a form similar to the single-fluid flow equation (3.35) in three space dimensions, the most apparent difference being the presence of leakage terms q_b and q_t. These terms account for any flux of fluid across the lower and upper confining surfaces, respectively, of the rock formation.

In steady flows, $\partial H / \partial t = 0$, and Eq. (3.50) reduces to the elliptic PDE

$$-\nabla \cdot (T\nabla H) = q, \tag{3.51}$$

where $q(\mathbf{x}) = q_b(\mathbf{x}) - q_t(\mathbf{x})$ denotes the net flux of water into the aquifer attributable to leakage.

3.6 Variational Forms for Steady Flow

Variational principles characterize certain solutions to PDEs as minimizing or maximizing some quantity. These principles arise frequently in functional analysis—the study of vector spaces of functions—and mathematical physics. Variational principles also play a prominent role in the theory of finite-element methods, which are among the most powerful numerical techniques used to solve the equations of fluid flow in porous media. This section introduces two variational formulations of single-fluid flows, using the steady areal flow equation (3.51) as an example. Both formulations figure in the convergence analysis of commonly used finite-element methods.

Consider a boundary-value problem of the form

$$\begin{aligned} -\nabla \cdot [T(\mathbf{x})\nabla H(\mathbf{x})] &= q(\mathbf{x}), \quad \text{for } \mathbf{x} \in \Omega; \\ H(\mathbf{x}) &= 0, \qquad \text{for } \mathbf{x} \in \partial\Omega. \end{aligned} \tag{3.52}$$

Assume that Ω is a two-dimensional spatial region amenable to the standard integral theorems of vector calculus, in particular having an orientable boundary $\partial\Omega$ on which there is a well-defined, outward-pointing, unit-length normal vector field $\mathbf{n}(\mathbf{x})$. Also, assume that the transmissivity T is a differentiable function of $\mathbf{x}$ that obeys bounds of the form $T_m \leqslant T(\mathbf{x}) \leqslant T_M$, for some positive constants T_m and T_M. Given a solution $H(\mathbf{x})$ to this problem, we seek real-valued expressions that the solution H minimizes.

3.6.1 Standard Variational Form

Recasting the problem (3.52) in a variational form requires three vector spaces of functions. Define

$$\mathcal{L}^2(\Omega) = \left\{ \varphi : \Omega \to \mathbb{R} \,\middle|\, \int_\Omega \varphi^2 \, dv \text{ exists} \right\},$$

$$\mathcal{H}^1(\Omega) = \left\{ \varphi \in \mathcal{L}^2(\Omega) \, \bigg| \int_\Omega \nabla\varphi \cdot \nabla\varphi \, dv \text{ exists} \right\},$$

$$\mathcal{H}_0^1(\Omega) = \left\{ \varphi \in \mathcal{H}^1(\Omega) \, \big| \, \varphi(\mathbf{x}) = 0, \text{ for all } \mathbf{x} \in \partial\Omega \right\}.$$

Exercise 3.13 *Assume that H is a solution to the boundary-value problem (3.52), so in particular H is twice differentiable. Use the product rule, in the form* $\nabla \cdot (\varphi\psi) = \varphi\nabla \cdot \psi + \nabla\varphi \cdot \psi$, *together with the divergence theorem, to show that*

$$\int_\Omega T\nabla H \cdot \nabla\varphi \, dv = \int_\Omega q\varphi \, dv, \quad \text{for all } \varphi \in \mathcal{H}_0^1(\Omega). \tag{3.53}$$

The **standard variational form** of the boundary-value problem (3.52) is to find a function $H \in \mathcal{H}_0^1(\Omega)$ such that Eq. (3.53) holds.

The variational form (3.53) is not equivalent to the original boundary-value problem (3.52). To be a solution of the PDE in Eq. (3.52), H must be twice differentiable, while Eq. (3.53) requires only that H have a square-integrable first derivative. Similar relaxations of smoothness requirements appear again in Chapters 4–6, in discussing weak solutions to PDEs.

Based on the results of Exercise 3.13, define an **energy functional** $F : \mathcal{H}_0^1(\Omega) \rightarrow \mathbb{R}$ as follows:

$$F(\varphi) = \frac{1}{2}\int_\Omega T\nabla\varphi \cdot \nabla\varphi \, dv - \int_\Omega q\varphi \, dv.$$

To establish the desired minimization principle, we show that H is a solution to the variational problem if and only if H minimizes the energy functional.

First assume that H solves the variational problem (3.53). It suffices to show that $F(\varphi) \geqslant F(H)$ for any $\varphi \in \mathcal{H}_0^1(\Omega)$. Writing $\varphi(\mathbf{x}) = H(\mathbf{x}) + \varepsilon(\mathbf{x})$ for some perturbation function $\varepsilon \in \mathcal{H}_0^1(\Omega)$, we find that

$$F(\varphi) = F(H + \varepsilon) = \frac{1}{2}\int_\Omega T\nabla(H + \varepsilon) \cdot \nabla(H + \varepsilon) \, dv - \int_\Omega q(H + \varepsilon) \, dv$$

$$= F(H) + \underbrace{\int_\Omega T\nabla\varepsilon \cdot \nabla\varepsilon \, dv}_{\text{(I)}} + \underbrace{\int_\Omega T\nabla H \cdot \nabla\varepsilon \, dv - \int_\Omega q\varepsilon \, dv}_{\text{(II)}}.$$

The term labeled (I) is nonnegative, since $T \geqslant T_m > 0$ and $\nabla\varepsilon \cdot \nabla\varepsilon \geqslant 0$, and the terms labeled (II) cancel each other by Eq. (3.53). It follows that H minimizes F.

Next assume that $F(H) \leqslant F(\varphi)$ for all functions $\varphi \in \mathcal{H}_0^1(\Omega)$. For any fixed choice of φ, define a function $G : \mathbb{R} \rightarrow \mathbb{R}$ by $G(s) = F(H + s\varphi)$. Since G is differentiable with a minimum value at $s = 0$, $G'(0) = 0$. But

$$G(s) = F(H) + s \left(\int_\Omega T\nabla H \cdot \nabla \varphi \, dv - \int_\Omega q\varphi \, dv \right)$$
$$+ \frac{1}{2}s^2 \int_\Omega T\nabla \varphi \cdot \nabla \varphi \, dv,$$

so differentiation with respect to s yields

$$0 = G'(0) = \int_\Omega T\nabla H \cdot \nabla \varphi \, dv - \int_\Omega q\varphi \, dv.$$

This result shows that the variational equation (3.53) holds for any $\varphi \in \mathcal{H}_0^1(\Omega)$.

3.6.2 Mixed Variational Form

The next two exercises introduce an alternative variational form that proves useful in many numerical approximations. Instead of starting with the second-order flow equation (3.51), let us retain Darcy's law and the steady mass balance as a pair of coupled first-order PDEs. Temporarily writing $\phi\mathbf{v} = \mathbf{u}$ for the filtration velocity, we have

$$K^{-1}\mathbf{u} + \nabla H = 0,$$
$$\nabla \cdot \mathbf{u} = q, \tag{3.54}$$

on the spatial domain Ω, with $H = 0$ on the boundary $\partial\Omega$.

Exercise 3.14 *Define the following vector spaces of functions:*

$$\mathcal{L}_0^2(\Omega) = \left\{ \psi \in \mathcal{L}^2(\Omega) \,\middle|\, \psi(0) = 0 \text{ on } \partial\Omega \right\},$$

$$\mathcal{H}(\text{div}, \Omega) = \left\{ \boldsymbol{\varphi} : \Omega \to \mathbb{R}^2 \,\middle|\, \int_\Omega \boldsymbol{\varphi} \cdot \boldsymbol{\varphi} \, dv \text{ exists and } \nabla \cdot \boldsymbol{\varphi} \in \mathcal{L}^2(\Omega) \right\}.$$

Let the pair $(\mathbf{u}, H)$ be a solution to the system (3.54). Show that

$$\int_\Omega K^{-1}\mathbf{u} \cdot \boldsymbol{\varphi} \, dv - \int_\Omega H\nabla \cdot \boldsymbol{\varphi} \, dv = 0, \qquad \text{for all } \boldsymbol{\varphi} \in \mathcal{H}(\text{div}, \Omega),$$

$$\int_\Omega \psi\nabla \cdot \mathbf{u} \, dv = \int_\Omega \psi q \, dv, \quad \text{for all } \psi \in \mathcal{L}_0^2(\Omega). \tag{3.55}$$

The **mixed variational form** of the first-order system (3.54) is to find $\mathbf{u} \in \mathcal{H}(\text{div}, \Omega)$ and $H \in \mathcal{L}_0^2(\Omega)$ such that Eqs. (3.55) hold. As in the variational form (3.53), the mixed variational form relaxes the smoothness requirements on the solution pair $(\mathbf{u}, H)$ that are implicit in the system (3.54).

The variational property associated with the system (3.55) has an interesting twist.

Exercise 3.15 *Using the vector spaces defined in Exercise 3.14, define the functional*

$$F(\boldsymbol{\varphi}, \psi) = \frac{1}{2} \int_{\Omega} K^{-1} \boldsymbol{\varphi} \cdot \boldsymbol{\varphi} \, dv - \int_{\Omega} \psi \nabla \cdot \boldsymbol{\varphi} \, dv + \int_{\Omega} \psi q \, dv.$$

Show that, if $(\mathbf{u}, H)$ *is a solution to the mixed variational form* (3.55), *then*

$$F(\mathbf{u}, \psi) \leqslant F(\mathbf{u}, H) \leqslant F(\boldsymbol{\varphi}, H), \text{for all } \boldsymbol{\varphi} \in \mathcal{H}(\text{div}, \Omega), \ \psi \in \mathcal{L}_0^2(\Omega). \qquad (3.56)$$

The inequalities (3.56) reveal that the solution $(\mathbf{u}, H)$ minimizes the functional $F(\boldsymbol{\varphi}, \psi)$ with respect to $\boldsymbol{\varphi}$ and maximizes F with respect to ψ. For this reason, we refer to the mixed variational form as a **saddle-point** problem.

The mixed variational form of the flow equation gives rise to **mixed finite-element methods**. These methods have significant virtues in the numerical solution of underground flow problems. In particular, careful choices of finite-element spaces yield approximate fluid velocities that are comparable in accuracy to the approximate solutions for H. For details, see [46] and [33, Chapter 3].

3.7 Flow in Anisotropic Porous Media

3.7.1 The Permeability Tensor

The assumption in Section 3.1 that permeability is a scalar function of position restricts the applicability of Eq. (3.8) to a narrow class of porous media. To see why, recall the formulation (3.24) of Darcy's law in terms of piezometric head:

$$\phi \mathbf{v} = -K \nabla H, \quad \text{where } K(\mathbf{x}) = \frac{\gamma g k(\mathbf{x})}{\mu}.$$

This formulation shows that the filtration velocity field $\phi(\mathbf{x})\mathbf{v}(\mathbf{x}, t)$ must be everywhere parallel to the vector field $\nabla H(\mathbf{x}, t)$, which we abbreviate as $\mathbf{G}(\mathbf{x}, t)$ in this section.

In many geologic porous media, the texture of the rock alters this relationship. For example, in the shale-streaked rock sample shown in Figure 3.12a, a piezometric head gradient $\mathbf{G}$ applied in a direction transverse to the geologic bedding planes may yield a macroscopic filtration velocity whose direction lies closer to the bedding planes. Figure 3.12b shows no such directional bias.

We accommodate this possibility by allowing the permeability to be a tensor-valued function $k(\mathbf{x})$. At each point $\mathbf{x}$ in the porous medium, this function defines a linear transformation such that

$$\phi(\mathbf{x}) \, \mathbf{v}(\mathbf{x}, t) = -\frac{\gamma g}{\mu} k(\mathbf{x}) \, \mathbf{G}(\mathbf{x}, t). \qquad (3.57)$$

Figure 3.12 Sandstone core samples. Sample (a) has dark shale streaks parallel to the geologic bedding planes. A pressure gradient transverse to the bedding planes in this rock may yield a filtration velocity that is not parallel to the pressure gradient. Sample (b) exhibits no obvious anisotropy.

(a)   (b)

The universe of linear transformations admits infinitely many cases in which $k(\mathbf{x})\,\mathbf{G}(\mathbf{x}, t)$ is not parallel to $\mathbf{G}(\mathbf{x}, t)$. In fact, this universe is too large, in a physically meaningful sense. Section 3.7.4 examines restrictions commonly imposed on the tensor field k. First, however, we review basic facts about tensors more generally.

3.7.2 Matrix Representations of the Permeability Tensor

Any choice of Cartesian coordinate system, defined by an orthonormal basis $\{\mathbf{e}_1, \mathbf{e}_2, \mathbf{e}_3\}$, yields a representation for the piezometric head gradient $\mathbf{G}(\mathbf{x}, t)$ as an $\mathbb{R}^3$-valued function:

$$\begin{bmatrix} G_1(\mathbf{x}, t) \\ G_2(\mathbf{x}, t) \\ G_3(\mathbf{x}, t) \end{bmatrix} = \begin{bmatrix} \partial H/\partial x_1 \\ \partial H/\partial x_2 \\ \partial H/\partial x_3 \end{bmatrix}(\mathbf{x}, t).$$

The functions $G_i = \mathbf{e}_i \cdot \mathbf{G} = \partial H/\partial x_i$, $i = 1, 2, 3$, are the **coordinate functions** of $\mathbf{G}$ with respect to the basis $\{\mathbf{e}_1, \mathbf{e}_2, \mathbf{e}_3\}$, giving

$$\mathbf{G} = \sum_{i=1}^{3} G_i \mathbf{e}_i.$$

The geometric properties of $\mathbf{G}$—its magnitude and direction at any point in space and time—remain unchanged under changes in coordinate systems. But the representation as an ordered triple in $\mathbb{R}^3$ changes, as discussed shortly.

The choice of orthonormal basis defines an analogous representation in $\mathbb{R}^3$ for the velocity field $\mathbf{v}(\mathbf{x}, t)$:

$$\begin{bmatrix} v_1(\mathbf{x}, t) \\ v_2(\mathbf{x}, t) \\ v_3(\mathbf{x}, t) \end{bmatrix}, \tag{3.58}$$

with

$$\mathbf{v} = \sum_{i=1}^{3} v_i \mathbf{e}_i.$$

Similarly, as discussed in Section 2.2, any tensor-valued function has a matrix representation with respect to the basis $\{\mathbf{e}_1, \mathbf{e}_2, \mathbf{e}_3\}$. Thus, the permeability tensor field $k(\mathbf{x})$ has matrix representation

$$\begin{bmatrix} k_{11}(\mathbf{x}) & k_{12}(\mathbf{x}) & k_{13}(\mathbf{x}) \\ k_{21}(\mathbf{x}) & k_{22}(\mathbf{x}) & k_{23}(\mathbf{x}) \\ k_{31}(\mathbf{x}) & k_{32}(\mathbf{x}) & k_{33}(\mathbf{x}) \end{bmatrix},$$

where $k_{ij}(\mathbf{x}) = \mathbf{e}_i \cdot [k(\mathbf{x})\mathbf{e}_j]$. The geometric action of $k(\mathbf{x})$ as a linear transformation remains independent of the choice of coordinate system. Using these representations for k, $\mathbf{G} = \nabla H$, and $\mathbf{v}$, we write the relationship (3.57) as a matrix equation:

$$\phi \begin{bmatrix} v_1 \\ v_2 \\ v_3 \end{bmatrix} = -\frac{\gamma g}{\mu} \begin{bmatrix} k_{11} & k_{12} & k_{13} \\ k_{21} & k_{22} & k_{23} \\ k_{31} & k_{32} & k_{33} \end{bmatrix} \begin{bmatrix} \partial H/\partial x_1 \\ \partial H/\partial x_2 \\ \partial H/\partial x_3 \end{bmatrix}.$$

This matrix form of Darcy's law is useful in computations.

To see how a change in Cartesian coordinate system affects the representations of vectors and tensors, consider a change from one orthonormal basis $\{\mathbf{e}_1, \mathbf{e}_2, \mathbf{e}_3\}$ to another $\{\hat{\mathbf{e}}_1, \hat{\mathbf{e}}_2, \hat{\mathbf{e}}_3\}$. With respect to these two bases, the vector field $\mathbf{G}$ has expansions

$$\mathbf{G} = \sum_{i=1}^{3} G_i \mathbf{e}_i = \sum_{i=1}^{3} \hat{G}_i \hat{\mathbf{e}}_i,$$

where the coefficients $\hat{G}_1, \hat{G}_2, \hat{G}_3$ are the coordinate functions of $\mathbf{G}$ with respect to the basis $\{\hat{\mathbf{e}}_1, \hat{\mathbf{e}}_2, \hat{\mathbf{e}}_3\}$. These coordinate functions may differ from G_1, G_2, G_3, even though the vector $\mathbf{G}(\mathbf{x})$ remains unchanged as a geometric entity at each point $\mathbf{x}$ in space. The following exercise shows how the coordinate functions change.

Exercise 3.16 *Prove the vector transformation rule*

$$\hat{G}_i = \sum_{j=1}^{3} Q_{ij} G_j,$$

*where the coefficients $Q_{ij} = \hat{\mathbf{e}}_i \cdot \mathbf{e}_j$ are the **direction cosines** between the basis elements $\{\mathbf{e}_1, \mathbf{e}_2, \mathbf{e}_3\}$ and $\{\hat{\mathbf{e}}_1, \hat{\mathbf{e}}_2, \hat{\mathbf{e}}_3\}$, illustrated in Figure 3.13. Show that*

$$G_i = \sum_{l=1}^{3} Q_{il}^{\mathsf{T}} \hat{G}_l,$$

where $Q_{il}^{\mathsf{T}} = Q_{li}$.

Figure 3.13 Direction cosine Q_{ij} between a vector $\mathbf{e}_j$ in one orthonormal basis and a vector $\hat{\mathbf{e}}_i$ in another.

An extension of the reasoning used in Exercise 3.16 shows how the entries of the matrix representation for the tensor-valued permeability function k change under this same change of coordinates.

Exercise 3.17 *If $k_{ij} = \mathbf{e}_i \cdot k\mathbf{e}_j$ and $\hat{k}_{ij} = \hat{\mathbf{e}}_i \cdot k\hat{\mathbf{e}}_j$, verify the tensor transformation rule*

$$k_{ij} = \sum_{l=1}^{3}\sum_{m=1}^{3} Q_{il}^{\mathsf{T}} \hat{k}_{lm} Q_{mj}, \qquad \hat{k}_{lm} = \sum_{i=1}^{3}\sum_{j=1}^{3} Q_{li}^{\mathsf{T}} k_{ij} Q_{jm}^{\mathsf{T}}.$$

(Hint: Look at the relationship

$$\mathbf{p} = k\mathbf{q} = \sum_{i=1}^{3} p_i \mathbf{e}_i = \sum_{i=1}^{3} \hat{p}_i \hat{\mathbf{e}}_j$$

for an arbitrary vector **q**.*)*

Occasionally one finds definitions of the term tensor based on these transformation rules. The perspective adopted here defines tensors as linear transformations whose geometric actions are invariant with respect to changes in the coordinate system. The transformation rules then follow from this invariance.

3.7.3 Isotropy and Homogeneity

When the permeability has a tensor character, Darcy's law becomes

$$\phi\mathbf{v} = -\frac{k}{\mu}(\nabla p - \gamma g \nabla Z).$$

One can accommodate the previous version, involving a scalar-valued permeability k, by writing $k = k\mathsf{I}$. Here I denotes the **identity tensor**, having the following matrix representation with respect to any orthonormal basis:

$$\begin{bmatrix} 1 & 0 & 0 \\ 0 & 1 & 0 \\ 0 & 0 & 1 \end{bmatrix}.$$

Whenever a tensor-valued material property, such as k, can be written as a (possibly spatially varying) scalar multiple of the identity tensor, the material is

Table 3.1 Isotropic, anisotropic, homogeneous, and inhomogeneous permeability tensors.

	Homogeneous	Inhomogeneous
Isotropic	$k\,\mathsf{I}$	$k(\mathbf{x})\,\mathsf{I}$
Anisotropic	k	$\mathsf{k}(\mathbf{x})$

isotropic with respect to that property. Otherwise, the material is **anisotropic**. In an isotropic material, the property may vary with spatial position, but it has no directional dependence.

A material property is **homogeneous** if its value is independent of spatial position $\mathbf{x}$. Otherwise, the material property is **inhomogeneous**. A tensor-valued property can be homogeneous and isotropic, homogeneous but anisotropic, isotropic but inhomogeneous, or anisotropic and inhomogeneous, as Table 3.1 illustrates.

3.7.4 Properties of the Permeability Tensor

Most engineers assume that the permeability tensor k has additional properties. To explore these properties in the simplest mathematical context, we examine a tensor version of Darcy's law written in terms of piezometric head:

$$\phi\mathbf{v} = -\mathsf{K}\nabla H.$$

Here, paralleling Eq. (3.30), $\mathsf{K}(\mathbf{x}) = (\gamma g/\mu)\mathsf{k}(\mathbf{x})$ denotes the tensor-valued hydraulic conductivity. Substituting this version of Darcy's law into the mass balance (3.34) yields the anisotropic single-fluid flow equation,

$$S_s\frac{\partial H}{\partial t} - \nabla\cdot(\mathsf{K}\nabla H) = 0.$$

The two most common additional properties attributed to the permeability tensor, cast in terms of the hydraulic conductivity, are as follows:

1. K is **positive semidefinite**, that is,

$$\mathbf{q}\cdot\mathsf{K}\mathbf{q} \geqslant 0, \quad \text{for all vectors } \mathbf{q}. \tag{3.59}$$

2. K is **symmetric**, that is,

$$\mathbf{p}\cdot(\mathsf{K}\mathbf{q}) = (\mathsf{K}\mathbf{p})\cdot\mathbf{q}, \quad \text{for all vectors } \mathbf{p}, \mathbf{q}. \tag{3.60}$$

(We temporarily suppress the dependence of K on spatial position $\mathbf{x}$, to keep the notation simple.) We now explore these two properties, beginning with a review of the relevant linear algebra.

Exercise 3.18   *The **transpose** of* K *is the tensor* K^T *defined by the condition that* $\mathbf{p} \cdot (K\mathbf{q}) = (K^T\mathbf{p}) \cdot \mathbf{q}$, *for all vectors* $\mathbf{p}, \mathbf{q}$. *Show that the* (i,j)*th entry of the matrix representation of* K^T *with respect to any orthonormal basis* $\{\mathbf{e}_1, \mathbf{e}_2, \mathbf{e}_3\}$ *is* $K_{ij}^T = K_{ji}$.

Exercise 3.19   *Without assuming that* K *is symmetric, consider the decomposition* $K = K_{symm} + K_{skew}$, *where*

$$K_{symm} = \frac{1}{2}(K + K^T), \quad K_{skew} = \frac{1}{2}(K - K^T).$$

Show that K_{symm} *is a symmetric tensor.*

The tensor K_{symm} is the **symmetric part** of K.

Exercise 3.20   *Show that the tensor* K_{skew} *obeys the identity*

$$\mathbf{p} \cdot K_{skew}\mathbf{q} = -(K_{skew}\mathbf{p}) \cdot \mathbf{q}, \quad \text{for all vectors } \mathbf{p}, \mathbf{q}. \tag{3.61}$$

In short, $K_{skew}^T = -K_{skew}$. Any tensor that obeys the condition (3.61) is a **skew tensor**. We call K_{skew} the **skew part** of K.

The assertion (3.59) that K is positive semidefinite reflects the constraint that a single fluid flowing in a porous medium never flows in a direction of increasing piezometric head. This prohibition ultimately arises from thermodynamic principles; see [4]. In symbols, we require $\nabla H \cdot \mathbf{v} \leqslant 0$, which, by Darcy's law, implies that $\nabla H \cdot K \nabla H \geqslant 0$. Since we have no *a priori* control over the direction or magnitude of the vector field ∇H, K must therefore satisfy the condition (3.59).

In many single-fluid flows, the hydraulic conductivity—and hence the permeability—is **positive definite**, meaning that

$$\mathbf{q} \cdot K\mathbf{q} \geqslant 0, \quad \text{for all vectors } \mathbf{q}, \text{ and}$$

$$\mathbf{q} \cdot K\mathbf{q} = 0, \quad \text{only if } \mathbf{q} = \mathbf{0}. \tag{3.62}$$

Physically, this assumption implies that the medium is permeable in all directions.

Exercise 3.21   *Show that* $\mathbf{q} \cdot K\mathbf{q} = \mathbf{q} \cdot K_{symm}\mathbf{q}$ *for any vector* $\mathbf{q}$. *Therefore the skew part* K_{skew} *has no effect on whether* K *is positive definite.*

Justifying the assertion (3.60) that K is symmetric is a more problematic task. We defer discussion of this question to the Section 3.7.5. For now, we observe that properties (3.60) and (3.62), taken together, have two important consequences. First, owing to symmetry, at any point $\mathbf{x}$ the tensor $K(\mathbf{x})$ has real eigenvalues $K_1(\mathbf{x}), K_2(\mathbf{x})$, and $K_3(\mathbf{x})$. The fact that K is positive definite ensures, in addition, that $K_1(\mathbf{x}), K_2(\mathbf{x})$, and $K_3(\mathbf{x})$ are all positive at every point $\mathbf{x}$.

Second, symmetry also guarantees that there exists, for each position $\mathbf{x}$, an orthonormal basis

$$\{\mathbf{p}_1(\mathbf{x}), \mathbf{p}_2(\mathbf{x}), \mathbf{p}_3(\mathbf{x})\}$$

for three-dimensional Euclidean space consisting of eigenvectors of $\mathsf{K}(\mathbf{x})$, that is, vectors for which

$$\mathsf{K}(\mathbf{x})\mathbf{p}_i(\mathbf{x}) = K_i(\mathbf{x})\mathbf{p}_i(\mathbf{x}) \quad \text{and} \quad \mathbf{p}_i(\mathbf{x}) \cdot \mathbf{p}_j(\mathbf{x}) = \begin{cases} 1, & \text{if } i = j, \\ 0, & \text{if } i \neq j. \end{cases}$$

We call these mutually orthogonal eigenvectors the **principal directions** of K at $\mathbf{x}$. The corresponding eigenvalues $K_1(\mathbf{x})$, $K_2(\mathbf{x})$, and $K_3(\mathbf{x})$ are the **principal hydraulic conductivities** at $\mathbf{x}$, which are the hydraulic conductivities at $\mathbf{x}$ of the porous medium in the directions of their respective eigenvectors. In the isotropic case, $K_1(\mathbf{x}) = K_2(\mathbf{x}) = K_3(\mathbf{x})$.

When the principal directions $\mathbf{p}_1(\mathbf{x}), \mathbf{p}_2(\mathbf{x}), \mathbf{p}_3(\mathbf{x})$ do not vary with spatial position $\mathbf{x}$, numerical modelers find it useful to adopt a Cartesian coordinate system aligned with these vectors [11]. Since the orthonormality of the basis implies

$$\mathbf{p}_i \cdot \mathsf{K}\mathbf{p}_j = \mathbf{p}_i \cdot K_j\mathbf{p}_j = \begin{cases} K_j, & \text{if } i = j, \\ 0, & \text{if } i \neq j, \end{cases} \tag{3.63}$$

Darcy's law in such a coordinate system—if it exists—has the matrix representation

$$\phi \begin{bmatrix} v_1 \\ v_2 \\ v_3 \end{bmatrix} = - \begin{bmatrix} K_1 & 0 & 0 \\ 0 & K_2 & 0 \\ 0 & 0 & K_3 \end{bmatrix} \begin{bmatrix} \partial H/\partial x_1 \\ \partial H/\partial x_2 \\ \partial H/\partial x_3 \end{bmatrix} = \begin{bmatrix} -K_1 \, \partial H/\partial x_1 \\ -K_2 \, \partial H/\partial x_2 \\ -K_3 \, \partial H/\partial x_3 \end{bmatrix}. \tag{3.64}$$

Equation (3.64) decouples the principal directions, in the sense that each scalar coordinate equation involves just one hydraulic conductivity and one partial derivative of piezometric head. In geologically simple settings, such as aquifers where the bedding planes are flat and the main effect of anisotropy is a difference between the vertical and horizontal conductivity, this decoupling greatly simplifies the computer coding required in a numerical simulator.

3.7.5 Is Permeability Symmetric?

We now examine more closely the question whether the hydraulic conductivity K (or, equivalently, the permeability k) is a symmetric tensor-valued function. The literature here can be frustrating, with published narratives ranging widely in explanatory power. In some expositions, the arguments for symmetry lack mathematical persuasiveness. In particular, as Exercise 3.21 shows, arguments based on properties of the quadratic form $\mathbf{p} \cdot \mathsf{K}\mathbf{p}$ (see [13, p. 13]) say nothing about K_{skew} and hence about whether $\mathsf{K} = \mathsf{K}_{\text{symm}}$.

In other, more defensible cases, symmetry emerges from spatial averaging procedures [94, 156]. Still other reasonable arguments appeal to thermodynamic principles beyond those derivable from the standard laws of mass balance, momentum balance, angular momentum balance, energy balance, and entropy inequality. For example, King et al. [88, 89] prove that the symmetry of k follows from minimizing the work required to move fluid through the porous medium or, equivalently, from more abstract principles advanced in 1931 by Norwegian-American chemist Lars Onsager [112].

However one views the matter, three observations have practical implications.

Observation 3.1   In some problems, only the symmetric part of the hydraulic conductivity tensor matters, as the following exercise shows.

Exercise 3.22   *Show that, if K is homogeneous but not necessarily symmetric, then*

$$\nabla \cdot (K \nabla H) = \nabla \cdot (K_{symm} \nabla H),$$

for any twice-differentiable function H.

In particular, the skew part of a spatially homogeneous hydraulic conductivity has no effect on the solution H to the flow equation in any boundary-value problem with pure Dirichlet boundary conditions (see Section 3.4). In such settings, when solving only for the piezometric head, we may as well use K_{symm} for computational convenience. This observation is irrelevant in transport problems (see Chapter 5), where we seek H for the purpose of calculating the filtration velocity. In particular, if K is not symmetric, then in general

$$\phi \mathbf{v} = -K \nabla H \neq -K_{symm} \nabla H.$$

Observation 3.2   Whatever standing one may confer on theoretical arguments, some mathematical modeling applications employ nonsymmetric permeability or hydraulic conductivity tensors. Among the most prominent of these applications are those that require upscaling K from the fine scale, where Darcy's law applies, to the coarser scales appropriate for spatial grid cells in field-scale numerical simulators. Upscaling methods typically yield nonsymmetric coarse-scale tensors, even if one starts with symmetric fine-scale tensors [88, 89, 157]. In such applications, one must either adopt numerical approximations that accommodate a nonsymmetric tensor or use an approximate tensor that is symmetric. The latter approach suffers from the disadvantage that it yields calculated flux directions different from the mathematically correct flux directions; see [165].

Observation 3.3   As convenient as the decoupling of gradients in Eq. (3.64) may be, it is not always available, even under the hypothesis that $K(\mathbf{x})$ is a symmetric tensor at each point **x**. If the principal directions $\mathbf{p}_1, \mathbf{p}_2, \mathbf{p}_3$ vary in space—as may happen, for example, when geologic forces have deformed the bedding planes of the rock formation—then a simplifying choice of Cartesian coordinate system that is globally aligned with the principal directions may not exist. Anisotropic problems of this nature typically allow no coordinate-wise decoupling of the type shown in Eq. (3.63), a fact that nullifies one of the most alluring consequences of symmetry in K.

4

Single-fluid Flow Problems

This chapter examines solutions to several classical problems involving single-fluid flows in porous media. These solutions reveal qualitative properties of solutions to more general problems. We begin by reviewing two linear, steady-state problems involving wells. We then examine an important transient problem from several perspectives, introducing mathematical concepts used in subsequent chapters. We close with a discussion of two mathematically related nonlinear problems involving single-fluid flows.

4.1 Steady Areal Flows with Wells

We examine two models of steady areal flows near wells. The first approach simplifies the physics in ways that, strictly speaking, are unrealistic in many applications. Nevertheless, the simplified model has an exact solution that relates pumping measurements to material properties of the rock formation. Solutions of this type provide entree into the topic of well-test analysis, a field that has evolved into a highly sophisticated set of methods for groundwater hydrologists [120, Chapter 6] and petroleum engineers [155, Chapter 3]. The second, more broadly applicable approach involves representing a well as an idealized point source or sink of fluid.

4.1.1 The Dupuit–Thiem Model

The first mathematical well model appeared in an 1863 treatise by French engineer Jules Dupuit [50, page 255], although German hydrologist Günter Thiem [146] and his father commonly get credit. Dupuit solved the areal flow equation near a well in a spatially uniform, confined groundwater aquifer. As shown below, one can compare the steady-state solution of this problem with field observations to calculate an aquifer's transmissivity.

The Mathematics of Fluid Flow Through Porous Media, First Edition. Myron B. Allen.
© 2021 John Wiley & Sons, Inc. Published 2021 by John Wiley & Sons, Inc.

The Dupuit–Thiem model rests on five simplifying assumptions. First, the fluid flow is areal and steady, with impermeable confining layers so that the leakage terms vanish. These assumptions reduce the areal flow equation (3.50) for the vertically averaged piezometric head to

$$\nabla \cdot (T\nabla H) = 0.$$

Second, the transmissivity T is spatially uniform and positive, so that the flow equation reduces further to the two-dimensional Laplace equation,

$$\nabla \cdot (\nabla H) = \nabla^2 H = 0. \tag{4.1}$$

Equation (4.1) is an elliptic PDE, typical of time-independent problems in mathematical physics and engineering.

Third, the flow is axisymmetric about the wellbore, so $H = H(r)$, where r denotes distance from the axis of the wellbore. The symmetry of the problem about the well suggests two changes of coordinates:

- Locate the origin **0** at the center of the wellbore's cross section.
- Convert to polar coordinates via the transformation

$$\Phi\left(\begin{bmatrix} r \\ \theta \end{bmatrix}\right) = \begin{bmatrix} r\cos\theta \\ r\sin\theta \end{bmatrix} = \begin{bmatrix} x_1 \\ x_2 \end{bmatrix},$$

as defined in Appendix B.

In these coordinates, Eq. (4.1) becomes

$$-\frac{1}{r}\frac{\partial}{\partial r}\left(r\frac{\partial H}{\partial r}\right) - \frac{1}{r^2}\frac{\partial^2 H}{\partial \theta^2} = 0;$$

see Eq. (B.3). Since $H = H(r)$, this PDE simplifies to the ordinary differential equation

$$\frac{1}{r}\frac{d}{dr}\left(r\frac{dH}{dr}\right) = 0.$$

Fourth, far from the well, at a distance r_e from its center, $H(r_e) = H_\infty$, a known value.

Fifth, the discharge rate at the well has a prescribed value Q, having dimension L^3T^{-1}. Darcy's law relates this rate to the derivative of the average piezometric head at the wellbore: If the wellbore is cylindrical with radius r_w and length l, then

$$2\pi r_w T\frac{dH}{dr}(r_w) = 2\pi r_w lK\frac{dH}{dr}(r_w) = Q.$$

Figure 4.1 shows the geometry.

These assumptions yield the following one-dimensional boundary-value problem for H:

$$\frac{d}{dr}\left(r\frac{dH}{dr}\right) = 0, \qquad r_w < r < r_e; \tag{4.2}$$

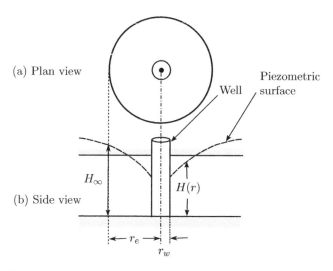

(a) Plan view

Well

Piezometric surface

H_∞

$H(r)$

(b) Side view

r_e

r_w

Figure 4.1 Geometry of the Dupuit–Thiem model.

$$H(r_e) = H_\infty, \qquad \frac{dH}{dr}(r_w) = \frac{Q}{2\pi r_w T}.$$

Exercise 4.1 *Show that the differential equation (4.2) has general solution* $H(r) = C_1 \log r + C_2$, *where* log *denotes the natural logarithm. Apply the boundary conditions to get*

$$H(r) = H_\infty + \frac{Q}{2\pi T} \log\left(\frac{r}{r_e}\right).$$

Thus, the piezometric head has a logarithmic singularity at the well. Show that, for any two radii $r_1, r_2 \in (r_w, r_e)$, *the **Thiem equation** holds:*

$$H(r_1) - H(r_2) = \frac{Q}{2\pi T} \log\left(\frac{r_1}{r_2}\right). \tag{4.3}$$

Equation (4.3) furnishes a method for calculating the transmissivity T using measured values of H. Define the **drawdown** as the difference $\bar{s}(r) = H_\infty - H(r)$ between the far-field vertically averaged piezometric head H_∞ and the vertically averaged piezometric head at distance r from the center of the wellbore, as shown in Figure 4.2. By Eq. (4.3),

$$\Delta\bar{s} = \bar{s}(r_1) - \bar{s}(r_2) = \frac{Q}{2\pi T} \log\left(\frac{r_1}{r_2}\right).$$

To use this relationship in a field study, one pumps water from an active well until the discharge rate reaches steady state, then measures the drawdown at a set of

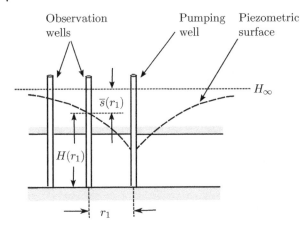

Figure 4.2 Definition of the drawdown $\bar{s}(r)$ at an observation well.

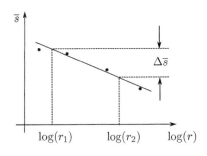

Figure 4.3 Calculation of transmissivity using drawdown data.

passive observation wells. By plotting $\bar{s}(r)$ versus $\log r$ and computing $\Delta\bar{s}$ on the best-fit line, as shown in Figure 4.3, we arrive at the value

$$T = \frac{Q}{2\pi\Delta\bar{s}}(\log r_1 - \log r_2).$$

This well test requires data from observation wells, which are typically expensive to drill if they do not already exist. Section 4.2 examines a well-testing method that requires data only from the well being tested.

4.1.2 Dirac δ Models

An alternative way to model a vertical well in more general areal flows is to treat the well as a point source or sink, represented as a Dirac δ distribution. In a heuristic sense, this approach represents a limiting case of the Dupuit–Thiem model as the wellbore radius $r_w \to 0$. It provides a convenient way to incorporate several wells in a single model. To develop the method, consider an areal flow with a single well, as discussed in Section 3.5, in a two-dimensional region Ω with boundary $\partial\Omega$. As in the Dupuit–Thiem model, we start by treating the wellbore as an internal

Figure 4.4 A two-dimensional region Ω with an external boundary $\partial\Omega_e$ and a wellbore modeled as an interior boundary $\partial\Omega_w$.

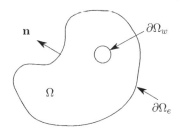

boundary $\partial\Omega_w$, denoting by $\partial\Omega_e$ the portion of the boundary that faces the exterior, as shown in Figure 4.4.

In the absence of leakage, the two-dimensional mass balance (3.46) for this region is

$$S\frac{\partial H}{\partial t} + \nabla \cdot (l\phi\mathbf{v}) = 0, \tag{4.4}$$

where l is the thickness of the formation, $\nabla\cdot$ is the two-dimensional divergence operator, and H and $\mathbf{v}$ are the vertically averaged piezometric head and two-dimensional velocity, respectively. Integrating this equation over Ω and applying the two-dimensional divergence theorem yields the integral mass balance:

$$\int_\Omega S\frac{\partial H}{\partial t}\, da = -\int_\Omega \nabla \cdot (l\phi\mathbf{v})\, da = -\int_{\partial\Omega} l\phi\mathbf{v} \cdot \mathbf{n}\, ds$$

$$= -\int_{\partial\Omega_e} l\phi\mathbf{v} \cdot \mathbf{n}\, ds - \int_{\partial\Omega_w} l\phi\mathbf{v} \cdot \mathbf{n}\, ds. \tag{4.5}$$

In this equation, the area integral on the left represents the net rate of change of fluid mass inside the region Ω, and both line integrals on the right trace the boundary $\partial\Omega$ in the positive sense, that is, keeping the interior of Ω on the left. The vector field $\mathbf{n}$ is the unit-length vector field pointing outward from the boundary.

Rewrite Eq. (4.5) as

$$\int_\Omega S\frac{\partial H}{\partial t}\, da + \int_{\partial\Omega_e} l\phi\mathbf{v} \cdot \mathbf{n}\, ds = Q. \tag{4.6}$$

In this version, the term

$$Q = -\int_{\partial\Omega_w} l\phi\mathbf{v} \cdot \mathbf{n}\, ds$$

represents the injection rate of the well, having dimension L^3T^{-1} and taking negative values when the well withdraws fluid from the formation.

In principle, the integral mass balance equation (4.6) allows for a detailed (but vertically averaged) description of the wellbore. In many field-scale settings, we forego detailed well models, reasoning that the cross-sectional area of the wellbore occupies a negligible fraction of the total area of the region Ω. Instead, we simplify

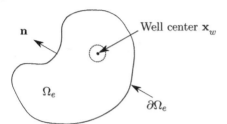

Figure 4.5 Two-dimensional region Ω_e used to model the region Ω in Figure 4.4 without the internal boundary $\partial\Omega_w$, shown here as a dotted-line circle.

the model, replacing Ω by a new region Ω_e, which is the union of Ω with the well, as depicted in Figure 4.5. To capture the effects of the well in this simplified region, we replace the mass balance (4.4) by a new equation,

$$S\frac{\partial H}{\partial t} + \nabla \cdot (l\phi\mathbf{v}) = Q\delta(\mathbf{x} - \mathbf{x}_w), \tag{4.7}$$

explained shortly. Here, $\mathbf{x}_w$ denotes the two-vector that coincides with the center of the wellbore.

The symbol δ stands for the two-dimensional **Dirac δ distribution**. This function-like entity has the following defining property: For any integrable function $\varphi(\mathbf{x})$ and any two-dimensional region $\mathcal{R}$,

$$\int_{\mathcal{R}} \varphi(\mathbf{x})\delta(\mathbf{x} - \mathbf{x}_w)\, da = \begin{cases} \varphi(\mathbf{x}_w), & \text{if } \mathbf{x}_w \in \mathcal{R}; \\ 0, & \text{otherwise.} \end{cases} \tag{4.8}$$

It is common to regard $\delta(\mathbf{x})$ as representing a **point source** having unit strength, since (i) it has nonzero effect only when integrated over regions $\mathcal{R}$ containing the point $\mathbf{x}_w$, and (ii) for such regions

$$\int_{\mathcal{R}} \delta(\mathbf{x} - \mathbf{x}_w)\, da = 1.$$

Strictly speaking, the right side of Eq. (4.7) does not arise from the mass balance law discussed in Chapter 2. Hence, the equation requires justification. By the defining property (4.8) and the divergence theorem, integration yields

$$\int_{\Omega_e} S\frac{\partial H}{\partial t}\, da + \int_{\partial\Omega_e} l\phi\mathbf{v} \cdot \mathbf{n}\, ds = \int_{\Omega_e} Q\delta(\mathbf{x} - \mathbf{x}_w)\, da$$
$$= \begin{cases} Q, & \text{if } \mathbf{x}_w \in \Omega_e, \\ 0, & \text{otherwise.} \end{cases} \tag{4.9}$$

Hence, we recover the integral mass balance (4.6). In short, by redefining the region of integration and including a compensating term involving the Dirac δ distribution, we account for the well's effects by treating it as a point source or sink, instead of explicitly treating the wellbore as part of the region's boundary.

4.1.3 Areal Flow in an Infinite Aquifer with One Well

As an example of a Dirac δ model, consider the steady, areal flow of a single fluid in a confined, horizontal aquifer having infinite extent and a single well centered at $\mathbf{x}_w$. The modified mass balance equation (4.7) reduces to the following time-independent form:

$$\nabla \cdot (l\phi\mathbf{v}) = Q\delta(\mathbf{x} - \mathbf{x}_w).$$

Substituting for $l\phi\mathbf{v}$ using the vertically averaged Darcy law

$$l\phi\mathbf{v} = -T\nabla H$$

(see Eq. (3.49)) yields

$$-\nabla \cdot (T\nabla H) = Q\delta(\mathbf{x} - \mathbf{x}_w).$$

In the simple case when T is constant, this equation reduces to a version of the **Poisson equation**,

$$-\nabla^2 H = \frac{Q}{T}\delta(\mathbf{x} - \mathbf{x}_w). \tag{4.10}$$

As with the Dupuit–Thiem model, the symmetry of this problem about the well location $\mathbf{x}_w$ suggests choosing the origin so that $\mathbf{x}_w = \mathbf{0}$ and converting to polar coordinates, defined in Appendix B. In these coordinates, Eq. (4.10) becomes

$$-\frac{1}{r}\frac{\partial}{\partial r}\left(r\frac{\partial H}{\partial r}\right) - \frac{1}{r^2}\frac{\partial^2 H}{\partial \theta^2} = \frac{Q}{T}\delta(r); \tag{4.11}$$

see Eq. (B.3). Since symmetry about the well implies that $H = H(r)$, the second term on the left side of Eq. (4.11) vanishes, and the governing equation further reduces to an ordinary differential equation on the interval $0 < r$:

$$\frac{1}{r}\frac{d}{dr}\left(r\frac{dH}{dr}\right) = 0. \tag{4.12}$$

To accommodate the Dirac δ distribution centered at $r = 0$, impose the following integral condition:

$$\lim_{r\to 0}\int_{D(r)} \nabla^2 H \, dv = -\frac{Q}{T}\lim_{r\to 0}\int_{D(r)} \delta(\mathbf{x}) \, dv = -\frac{Q}{T}, \tag{4.13}$$

where $D(r)$ denotes the disk of radius r centered at the wellbore.

The differential equation (4.12) has general solution

$$H(r) = C_1 \log r + C_2,$$

where C_1 and C_2 are any constants. The integral condition (4.13) suffices to determine C_1 as follows. By the two-dimensional divergence theorem, the area integral of $\nabla^2 H$ over the disk $D(r)$ reduces to a path integral around the bounding circle

$\partial D(r)$. Parametrizing this circle as $\zeta(\theta) = (r\cos\theta, r\sin\theta)$ for $0 \le \theta \le 2\pi$ and using the expression (B.2) for the two-dimensional gradient operator gives

$$
\begin{aligned}
-\frac{Q}{T} &= \lim_{r\to 0}\int_{D(r)} \nabla^2 H \, da = \lim_{r\to 0}\int_{D(r)} \nabla \cdot \nabla H \, da \\
&= \lim_{r\to 0}\int_{\partial D(r)} \nabla H \cdot \mathbf{n} \, ds \\
&= \lim_{r\to 0}\int_0^{2\pi} \left(\cos\theta \frac{dH}{dr}, \sin\theta \frac{dH}{dr}\right) \cdot (\cos\theta, \sin\theta) \, \|\zeta(\theta)\| \, d\theta \\
&= -2\pi \lim_{r\to 0} r\frac{dH}{dr},
\end{aligned}
\tag{4.14}
$$

the last identity following from the facts that $\|\zeta(\theta)\| = r$ and dH/dr is constant on the circle of radius r.

Exercise 4.2 *Deduce the following equations from the condition* (4.14):

$$
C_1 = \lim_{r\to 0} r\frac{dH}{dr} = -\frac{Q}{2\pi T}, \quad C_2 = H(1).
$$

Since H represents the vertically averaged piezometric head, which is a height above some arbitrarily chosen datum, we may choose the datum to guarantee that $C_2 = 0$ without affecting the fluid velocity field. Doing so yields the following solution to the Dirac δ well model:

$$
H(r) = -\frac{Q}{2\pi T} \log r. \tag{4.15}
$$

This solution shows that the vertically averaged piezometric head H varies smoothly except for a logarithmic singularity at the well. When $Q < 0$, corresponding to fluid withdrawal from the aquifer, $H(r) \to -\infty$ as $r \to 0$.

Although Eq. (4.15) is the solution to a highly idealized problem, it has a significant practical use. Knowing the structure of the singularity near a well helps improve the quality of numerical solutions to more realistic problems for which no closed-form solutions are available. Consider, for example, a steady areal flow problem having the form

$$
-\nabla \cdot [T(\mathbf{x})\nabla H] = Q\delta(\mathbf{x}), \tag{4.16}
$$

in which the transmissivity varies in space. Numerical methods, such as finite-difference or finite-element methods, typically produce highly accurate approximate solutions in problems for which the exact solution is **regular**, that is, possesses no singularities. However, the accuracy deteriorates near singularities. A technique called **singularity removal** [98] incorporates the structure of the singularity into the approximate solution *a priori*, allowing the numerical method to solve for the regular part of the solution.

To illustrate the method, first decompose $H = H_r + H_s$, where H_s denotes the singular part of the unknown solution and the remainder, H_r, is regular. Specifically, let H_s be the exact solution to a problem of the form

$$-\nabla^2 H_s = \frac{Q}{T_{avg}} \delta(\mathbf{x}), \tag{4.17}$$

calculating the constant T_{avg} as the average transmissivity in some region containing the well. For example, one can take

$$T_{avg} = \frac{1}{\pi R^2} \int_0^R \int_0^{2\pi} T(r, \theta) \, d\theta \, dr,$$

where $R > 0$ is the radius of a disk inside which the values of T are representative of values affecting the well's behavior. By Eq. (4.15),

$$H_s(r) = -\frac{Q}{2\pi T_{avg}} \log r. \tag{4.18}$$

Subtracting Eq. (4.17) from Eq. (4.16) and moving all known, computable terms to the right side yields

$$-\nabla \cdot (T \nabla H_r) = \underbrace{\nabla \cdot [(T - T_{avg}) \nabla H_s]}_{\text{known}}. \tag{4.19}$$

By calculating the right side using Eq. (4.18), we can solve Eq. (4.19) for the regular part H_r of the solution—a task at which classical numerical methods, such as finite-difference or finite-element methods, typically excel.

4.2 The Theis Model for Transient Flows

4.2.1 Model Formulation

In 1935, American engineer Charles V. Theis [145] presented an alternative method for determining aquifer parameters, based on the time-dependent hydraulics of a single well. Theis examined areal flow near a well in a horizontal, confined aquifer that has no leakage and constant storativity S and transmissivity T, both of which are positive. If the flow is axisymmetric, then by the areal flow equation (3.50) the vertically averaged piezometric head H obeys a version of the two-dimensional heat equation,

$$\nabla^2 H = \frac{S}{T} \frac{\partial H}{\partial t},$$

or, in polar coordinates (see Section B.1),

$$\frac{1}{r} \frac{\partial}{\partial r} \left(r \frac{\partial H}{\partial r} \right) = \frac{S}{T} \frac{\partial H}{\partial t}. \tag{4.20}$$

Equation (4.20) is a linear, parabolic PDE. Equations of this type often govern time-dependent processes and tend to smooth the initial conditions. Section 4.2.4 justifies this intuition for Eq. (4.20). However, as a nonlinear problem examined in Section 4.3 illustrates, the heuristic requires caution.

Consider an aquifer in which the vertically averaged piezometric head H has a spatially uniform initial value H_0, and assume that $H \to H_0$ in the far field, as $r \to \infty$. If we start pumping the well at time $t = 0$ with constant discharge rate Q, then H obeys Eq. (4.20) with the following initial and boundary conditions:

$$H(r, 0) = H_0,$$
$$\lim_{r \to \infty} H(r, t) = H_0, \tag{4.21}$$
$$\lim_{r \to 0} 2\pi r T \frac{\partial H}{\partial r}(r, t) = Q.$$

As with the Dupuit–Thiem model, the boundary condition as $r \to 0$ arises from Darcy's law.

4.2.2 Dimensional Analysis of the Theis Model

Dimensional analysis—the technique used to infer the form of the solution to the Stokes problem in Section 2.4—also reveals information about the functional dependencies in solutions to the Theis model. We begin by converting Eq. (4.20) to a form in which the unknown solution is dimensionless.

Exercise 4.3 *Define a dimensionless piezometric head*

$$u(r, t) = \frac{H - H_0}{Q/(2\pi T)},$$

and show that Eq. (4.20), together with the initial and boundary conditions (4.21), reduce to the following problem:

$$\frac{\partial^2 u}{\partial r^2} + \frac{1}{r}\frac{\partial u}{\partial r} - \frac{S}{T}\frac{\partial u}{\partial t} = 0, \qquad\qquad r, t > 0; \tag{4.22}$$
$$u(r, 0) = 0, \qquad\qquad r > 0; \tag{4.23}$$
$$\lim_{r \to 0} r \frac{\partial u}{\partial r}(r, t) = 1, \qquad\qquad t > 0; \tag{4.24}$$
$$\lim_{r \to \infty} u(r, t) = 0, \qquad\qquad t > 0. \tag{4.25}$$

The initial-boundary-value problem derived in Exercise 4.3 involves only the variables r, t, T, S, and u. Any solution u to the problem therefore defines a relationship among these variables, which we write conceptually as follows:

$$\varphi(r, t, T, S, u) = 0. \tag{4.26}$$

The variables in this relationship have physical dimensions L, T, L^2T^{-1}, 1, and 1, respectively.

Appealing to the Buckingham Pi theorem, reviewed in Appendix C, we seek an equivalent relationship in terms of a possibly smaller set of dimensionless variables. Any dimensionless variable Π for this problem has the form of a product of powers,

$$\Pi = r^{n_1} t^{n_2} T^{n_3} S^{n_4} u^{n_5},$$

with the exponents $n_1, \dots, n_5$ to be determined. Substituting the dimensions of the variables r, t, T, S, u and insisting that Π have dimension 1 yields

$$L^{n_1} T^{n_2} (L^2 T^{-1})^{n_3} 1^{n_4} 1^{n_5} = L^{n_1 + 2n_3} T^{n_2 - n_3} = 1.$$

The last identity holds only if the exponents of L and T vanish, a condition that implies the following homogeneous linear system for $n_1, \dots, n_5$:

$$\begin{bmatrix} 1 & 0 & 2 & 0 & 0 \\ 0 & 1 & -1 & 0 & 0 \end{bmatrix} \begin{bmatrix} n_1 \\ n_2 \\ n_3 \\ n_4 \\ n_5 \end{bmatrix} = \begin{bmatrix} 0 \\ 0 \end{bmatrix}.$$

This homogeneous, linear system is in row-reduced form. It shows that n_3, n_4, and n_5 are free variables, $n_2 = n_3$, and $n_1 = -2n_3$. We calculate three independent dimensionless variables by choosing three linearly independent vectors (n_3, n_4, n_5):

When $(n_3, n_4, n_5) = (1, 0, 0)$, $n_1 = -2, n_2 = 1$; $\quad \Pi_1 = \dfrac{tT}{r^2}$.

When $(n_3, n_4, n_5) = (0, 1, 0)$, $n_1 = 0, n_2 = 0$; $\quad \Pi_2 = S$.

When $(n_3, n_4, n_5) = (0, 0, 1)$, $n_1 = 0, n_2 = 0$; $\quad \Pi_3 = u$.

Therefore, the dimensional relationship (4.26) reduces to a relationship of the following form, involving only dimensionless variables:

$$\Phi\left(\frac{tT}{r^2}, S, u\right) = 0. \tag{4.27}$$

If a solution u exists, Eq. (4.27) shows that $u = U(S, r^2/(Tt))$, for some implicitly defined function U. Treating S and T as given positive constants and r, t as the independent variables in this problem, we write

$$u(r, t) = U\left(\frac{r^2}{t}\right). \tag{4.28}$$

The conclusion (4.28) reveals an important property of the solution $u(r, t)$ to the Theis problem: Except for a scaling factor, it retains its shape, defined by the graph of the function U, as t increases.

This observation merits a closer look. We call a function $u(r, t)$ **self-similar** if there exist exponents a, b such that

$$u(\varepsilon r, \varepsilon^a t) = \varepsilon^b u(r, t), \tag{4.29}$$

for all positive values of ε. We call the arbitrary positive constant ε a **scaling parameter**.

Exercise 4.4 *As a simple example, show that the function* $\exp(-Cr^2/t)$ *is self-similar for any positive constant C, with exponents $a = 2$ and $b = 0$. To see how the function retains its shape as t increases, plot the function for several fixed values of t.*

Equation (4.28) shows that the solution to the Theis problem given by Eqs. (4.22) through (4.25) is self-similar, with $a = 2$ and $b = 0$, since

$$u(\varepsilon r, \varepsilon^2 t) = U\left(\frac{\varepsilon^2 r^2}{\varepsilon^2 t}\right) = U\left(\frac{r^2}{t}\right) = \varepsilon^0 u(r, t).$$

So far, dimensional analysis has yielded only qualitative properties of the solution. An additional remark sets the stage for Section 4.2.4, which exploits self-similarity to develop an explicit formula for $u(r, t)$. For any fixed values of a and b, the mapping

$$g_\varepsilon(r, t, u) = (\varepsilon r, \varepsilon^a t, \varepsilon^b u), \quad \varepsilon > 0, \tag{4.30}$$

arising from the self-similarity equation (4.29), defines a one-parameter family of **stretching transformations**.

Exercise 4.5 *Fix the values of a and b, and denote by G the set of all stretching transformations having the form (4.30). Consider the binary operation defined by composing elements of G:*

$$(g_\delta \circ g_\varepsilon)(r, t, u) = g_\delta(g_\varepsilon(r, t, u)), \quad \delta, \varepsilon > 0.$$

Show that G has the following properties:

1. *G is closed under $\circ$, that is, for any stretching transformations $g_\delta, g_\varepsilon \in G$, the composition $g_\delta \circ g_\varepsilon$ belongs to G.*
2. *Composition is associative, that is, $g_\alpha \circ (g_\delta \circ g_\varepsilon) = (g_\alpha \circ g_\delta) \circ g_\varepsilon$ for all positive stretching parameters $\alpha, \delta, \varepsilon$.*
3. *G has an identity element g_1, for which $g_1(r, t, u) = (r, t, u)$.*
4. *Every stretching transformation g_ε in G has an inverse: There exists a stretching transformation g_ε^{-1} in G such that $g_\varepsilon^{-1} \circ g_\varepsilon = g_1 = g_\varepsilon \circ g_\varepsilon^{-1}$.*

Properties 1 through 4 show that G is a **group** under the operation of composition. We exploit this observation shortly.

4.2.3 The Theis Drawdown Solution

The linear initial-boundary-value problem given by Eqs. (4.20) and (4.21) admits several solution methods. An interesting approach, which extends to nonlinear PDEs analyzed in later sections, is the method of self-similar solutions, foreshadowed in Section 4.2.2 and pursued in detail in Section 4.2.4.

For now we simply state the result:

$$H(r,t) = H_0 + \frac{Q}{4\pi T} \operatorname{Ei}\left(-\frac{S}{4T}\frac{r^2}{t}\right), \tag{4.31}$$

where

$$\operatorname{Ei}(\xi) = -\int_{-\xi}^{\infty} \frac{e^{-y}}{y}\, dy \tag{4.32}$$

denotes the **exponential integral** function. The functional dependencies in this solution are consistent with the self-similar form (4.28), derived using dimensional analysis. Writing the solution (4.31) in terms of drawdown yields

$$\bar{s}(r,t) = H_0 - H(r,t) = -\frac{Q}{4\pi T}\operatorname{Ei}\left(-\frac{S}{4T}\frac{r^2}{t}\right), \quad \text{for } t > 0.$$

Figure 4.6 shows graphs of H and the drawdown at various times. Two observations are worth noting:

1. The drawdown $\bar{s}(r,t)$ is continuously differentiable to all orders.
2. For any $t > 0$, $\bar{s}(r,t)$ has nonzero values for all $r \geq 0$.

These two properties typify solutions to the heat equation: They are smooth after the initial time, and nonzero initial data instantaneously yield nonzero (but rapidly decaying) solution values throughout the spatial domain. In this sense, solutions to the linear heat equation propagate with infinite speed, a property that contrasts with solutions to some closely related nonlinear problems, as Section 4.3 explores.

In 1946, American hydrologists Hilton H. Cooper and Charles E. Jacob [39] proposed using an approximation for Ei $(-\xi)$ to turn the solution (4.31) into a practical

Figure 4.6 Graphs of the Theis solution for vertically averaged piezometric head at several time levels, showing the drawdown at time $t_1 > 0$.

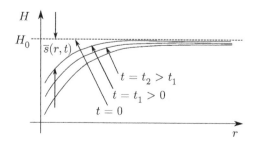

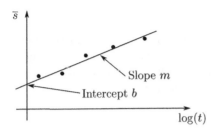

Figure 4.7 Hypothetical plot of drawdown versus log(t) for the Theis method.

well test to determine T and S. Citing the series expansion (see [7, Section 3.3])

$$\text{Ei}(\xi) = \gamma + \log|\xi| + \sum_{j=1}^{\infty} \frac{\xi^j}{jj!}, \quad \xi \neq 0,$$

where $\gamma \simeq 0.577216$ denotes Euler's constant and log stands for the natural logarithm, they adopted the approximation

$$\text{Ei}(-\xi) \simeq 0.577216 + \log|\xi|, \quad \text{for } \xi = \frac{r^2 S}{4Tt} < 0.01.$$

The **Cooper–Jacob approximate solution** is therefore

$$\bar{s}(r,t) \simeq -\frac{Q}{4\pi T}\left(0.577216 + \log\frac{r^2 S}{4Tt}\right) \simeq \frac{Q}{4\pi T}\log\left(\frac{2.25Tt}{r^2 S}\right),$$

valid for small values of r and large values of t.

Exercise 4.6 *Develop a method for computing T and S using this equation and drawdown data from a single well, as plotted in Figure 4.7.*

4.2.4 Solving the Theis Model via Similarity Methods

We now analyze the Theis model more closely, solving the linear initial-boundary-value problem given by Eqs. (4.22) through (4.25). We use the method of self-similar solutions, mentioned briefly in Section 4.2.2. While the problem admits several other solution techniques, the approach pursued here furnishes a productive line of attack for certain nonlinear problems, as subsequent sections illustrate.

The method exploits a **symmetry** of the PDE (4.20). By a symmetry we mean a group of transformations of the independent variables x, t, and the unknown solution u that leave the equation invariant, in a sense to be made precise shortly. The particular type of symmetry that we seek for Eq. (4.22) is a group of stretching transformations, as introduced in Exercise 4.5. For fixed real exponents a and b, any transformation of this type has the form $g_\varepsilon(r,t,u) = (\varepsilon r, \varepsilon^a t, \varepsilon^b u)$ given in Eq. (4.30), for some positive scaling parameter ε.

To describe the invariance condition, call

$$\xi = \varepsilon r, \quad \tau = \varepsilon^a t, \quad \eta(\xi, \tau) = \varepsilon^b u(r, t). \tag{4.33}$$

For Eq. (4.22), **invariance** requires that

$$\frac{\partial^2 \eta}{\partial \xi^2} + \frac{1}{\xi}\frac{\partial \eta}{\partial \xi} - \frac{S}{T}\frac{\partial \eta}{\partial \tau} = C \times \left(\frac{\partial^2 u}{\partial r^2} + \frac{1}{r}\frac{\partial u}{\partial r} - \frac{S}{T}\frac{\partial u}{\partial t} \right), \tag{4.34}$$

for some nonzero constant C. In other words, $u(r, t)$ solves the PDE in the original independent variables (r, t) if and only if $\eta(\xi, \tau)$ solves the PDE in the stretched independent variables $(\xi, \tau) = (\varepsilon r, \varepsilon^a t)$.

If such a symmetry exists, then—as we show below—two useful conclusions follow. First, the solution has the form

$$u(r, t) = t^{b/a} U(\zeta), \tag{4.35}$$

for some function U of a single variable $\zeta(r, t)$. As time progresses, a solution having the form (4.35) retains the shape $U(\zeta)$, stretched by a purely time-dependent factor $t^{-b/a}$. The graphs of piezometric head in Figure 4.6 exhibit this important geometric property. We show below that a function of the form (4.35) is self-similar, as formally defined by the condition (4.29). Second, we can determine the function U by solving an ordinary differential equation.

The derivation proceeds in four steps:

1. Show that if a symmetry for the PDE consisting of scaling transformations (4.30) exists, then the PDE has a solution form (4.35).
2. Verify that there exists such a symmetry for the governing PDE (4.20).
3. Derive the ordinary differential equation for the function $U(\zeta)$.
4. Solve for $U(\zeta)$ and hence for $u(r, t)$.

The first step is the longest. It rests on the observation that any smooth solution $u(r, t)$ of the PDE defines a surface in (r, t, u)-space having the form $\varphi(r, t, u) = 0$. The invariance condition (4.34) implies that

$$\varphi(g_\varepsilon(r, t, u)) = \varphi(\varepsilon r, \varepsilon^a t, \varepsilon^b u) = \varphi(r, t, u)$$

for all $\varepsilon > 0$. Therefore, $\varphi(\varepsilon r, \varepsilon^a t, \varepsilon^b u) = 0$, and

$$
\begin{aligned}
0 &= \frac{d}{d\varepsilon} \varphi(\varepsilon r, \varepsilon^a t, \varepsilon^b u) \\
&= \frac{d(\varepsilon r)}{d\varepsilon} \partial_1 \varphi + \frac{d(\varepsilon^a t)}{d\varepsilon} \partial_2 \varphi + \frac{d(\varepsilon^b u)}{d\varepsilon} \partial_3 \varphi \\
&= r\frac{\partial \varphi}{\partial r} + a\varepsilon^{a-1} t\frac{\partial \varphi}{\partial t} + b\varepsilon^{b-1} u\frac{\partial \varphi}{\partial u},
\end{aligned}
$$

by the chain rule. In particular, this equation holds when $\varepsilon = 1$, and hence φ obeys the following first-order PDE:

$$0 = r\frac{\partial \varphi}{\partial r} + at\frac{\partial \varphi}{\partial t} + bu\frac{\partial \varphi}{\partial u}. \tag{4.36}$$

Now consider any continuously differentiable path $(r(s), t(s), u(s))$ in (r, t, u)-space. The chain rule requires that φ change along this path according to the equation

$$\frac{d\varphi}{ds} = \frac{dr}{ds}\frac{\partial\varphi}{\partial r} + \frac{dt}{ds}\frac{\partial\varphi}{\partial t} + \frac{du}{ds}\frac{\partial\varphi}{\partial u}. \tag{4.37}$$

If we restrict attention to paths for which

$$\frac{dr}{ds} = r, \quad \frac{dt}{ds} = at, \quad \frac{du}{ds} = bu, \tag{4.38}$$

then consistency between Eqs. (4.36) and (4.37) requires that

$$\frac{d\varphi}{ds} = 0.$$

Thus, φ is constant along all paths in (r, t, u)-space that satisfy Eqs. (4.38). We call these paths **characteristic curves**. Rearranging the differential equations (4.38) shows that φ is constant along paths for which

$$\frac{dr}{dt} = \frac{r}{at} \quad \text{and} \quad \frac{du}{dt} = \frac{bu}{at}. \tag{4.39}$$

Exercise 4.7 *Integrate the differential equations in* (4.39) *with respect to* t *to obtain solutions of the form*

$$r = C_1 t^{1/a}, \quad u = C_2 t^{b/a},$$

where C_1, C_2 *are arbitrary constants.*

The results of Exercise 4.7 show that φ is constant along paths for which $r/t^{1/a}$ and $u/t^{b/a}$ are constant, that is $\varphi(r, t, u) = \Phi(r/t^{1/a}, u/t^{b/a})$, for some function Φ. On the solution surface, $\varphi(r, t, u) = 0$, so

$$\Phi\left(\frac{r}{t^{1/a}}, \frac{u}{t^{b/a}}\right) = 0. \tag{4.40}$$

Equation (4.40) implicitly defines a relationship $u/t^{b/a} = U(r/t^{1/a})$, which we rearrange to conclude that

$$u = t^{b/a}U(\zeta), \quad \text{where } \zeta = \frac{r}{t^{1/a}},$$

as claimed.

The second step shows that there exists a symmetry for the PDE consisting of stretching transformations (4.30).

Exercise 4.8 *Using the definition* (4.33) *of the stretching transformation, show that*

$$\frac{\partial\eta}{\partial\tau} = \varepsilon^{b-a}\frac{\partial u}{\partial t}; \quad \frac{1}{\xi}\frac{\partial\eta}{\partial\xi} = \varepsilon^{b-2}\frac{1}{r}\frac{\partial u}{\partial r}; \quad \frac{\partial^2\eta}{\partial\xi^2} = \varepsilon^{b-2}\frac{\partial^2 u}{\partial r^2}.$$

Exercise 4.8 reveals that the invariance condition (4.34) holds if $a = 2$. Also, the scaling condition $\eta(\varepsilon r, \varepsilon^2 t) = \varepsilon^b u(r, t)$ holds for all $\varepsilon > 0$, with the exponent b still undetermined. Thus, $\zeta = rt^{-1/2}$, but it is equivalent (and simpler) to take

$$u = t^{b/2}U(\zeta), \quad \text{where } \zeta = \frac{r^2}{t}. \tag{4.41}$$

For the third step, we show that the representation (4.41) for $u(r, t)$ reduces the PDE (4.22) to an ordinary differential equation for $U(\zeta)$.

Exercise 4.9 *Calculate $\partial u/\partial t$, $\partial u/\partial r$, and $\partial^2 u/\partial r^2$ in terms of $U(\zeta)$ and show that the PDE (4.22) implies the following ordinary differential equation for U:*

$$\zeta U'' + (1 + \kappa\zeta)U' - \frac{\kappa b}{2}U = 0, \quad \text{where } \kappa = \frac{S}{4T}. \tag{4.42}$$

The fourth step—solving Eq. (4.42)—requires translating the initial and boundary conditions for $u(r, t)$, given in Eqs. (4.23), (4.24), and (4.25), to conditions on $U(\zeta)$.

Exercise 4.10 *From the outer boundary condition (4.25), derive the condition*

$$\lim_{\zeta \to \infty} U(\zeta) = 0.$$

Show that the inner boundary condition (4.24) yields the following condition:

$$\lim_{r \to 0} r\frac{\partial u}{\partial r} = \lim_{\zeta \to 0} 2t^{b/2}\zeta U'(\zeta) = 1. \tag{4.43}$$

For the limit (4.43) to be constant for all values of $t > 0$, b must vanish. Thus

$$u(\varepsilon r, \varepsilon^2 t) = U(\zeta) = \varepsilon^0 u(r, t),$$

showing that $u(r, t)$ is self-similar by the definition (4.29). In addition, the ordinary differential equation (4.42) simplifies, and the function $U(\zeta)$ must solve the following boundary-value problem:

$$U'' + \left(\frac{1}{\zeta} + \kappa\right)U', \tag{4.44}$$

$$\lim_{\zeta \to 0} \zeta U' = \frac{1}{2}, \tag{4.45}$$

$$\lim_{\zeta \to \infty} U = 0. \tag{4.46}$$

Integrating Eq. (4.44) once using the integrating factor

$$\exp\int \left(\frac{1}{\zeta} + \kappa\right) d\zeta = \zeta\exp(\kappa\zeta)$$

and the boundary condition (4.45) yields

$$U' = \frac{1}{2\zeta} \exp(-\kappa\zeta).$$

Integrating again and using the boundary condition (4.46) give

$$U(\zeta) = -\frac{1}{2} \int_\zeta^\infty \frac{e^{-\kappa y}}{y} \, dy = -\frac{1}{2} \int_{\kappa\zeta}^\infty \frac{e^{-y}}{y} \, dy = \frac{1}{2} \mathrm{Ei}(-\kappa\zeta),$$

where Ei denotes the exponential integral defined in Eq. (4.32). Remembering that $\zeta = r^2/t$ and $\kappa = S/(4T)$, we have

$$\frac{H - H_0}{Q/(2\pi T)} = u(r,t) = U\left(\frac{r^2}{t}\right) = \frac{1}{2} \mathrm{Ei}\left(-\frac{S}{4T}\frac{r^2}{t}\right),$$

which is equivalent to the Theis solution (4.31).

The method of self-similar solutions has limitations, even in cases when the PDE is invariant under a group of stretching transformations. For example, it may not be possible to solve the ordinary differential equation for $U(\zeta)$ by analytic methods. More significantly, not all initial-boundary-value problems have self-similar solutions. Finite spatial domains, for example, tend to thwart the search for self-similar solutions that satisfy given boundary conditions.

This limitation notwithstanding, Russian mathematician Grigory I. Barenblatt and Russian physicist Yakov B. Zel'dovich [14] argued that self-similar solutions reveal important structural aspects of PDEs. In this view, far from being special cases, self-similar solutions describe the **intermediate asymptotic** behavior of more general solutions, that is, the behavior of these solutions far from spatial and temporal boundaries.

We encounter the method of self-similar solutions again in Section 4.3, which investigates a nonlinear single-fluid flow problem; in Section 5.2, in connection with the linear transport equation; and in Section 6.7, in the analysis of nonlinear three-fluid flows in porous media.

4.3 Boussinesq and Porous Medium Equations

So far, all of the problems that we have considered involve a single fluid flowing in a confined porous medium. In these settings, the fluid—typically water—completely fills the pore space, and hence, the geometry of the medium defines the geometry of the problem. Mathematical complications arise when a single fluid does not fill the pore space, as in an **unconfined** or **phreatic aquifer**, where air is also present. (The word phreatic comes from the ancient Greek word φρέαρ for water well.) In unconfined aquifers, the top of the porous

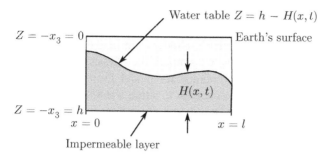

Figure 4.8 Schematic diagram of a vertical slice through an unconfined aquifer, showing the water table $Z = h - H(x, t)$.

medium is open to the atmosphere, and water occupies a subregion of the aquifer whose vertical extent can change as fluid flows through the medium.

Figure 4.8 shows a vertical slice through such an aquifer. For simplicity, assume that the aquifer occupies a spatial region defined by the inequalities

$$0 \leq x_1 \leq l,$$
$$0 \leq x_2 \leq w,$$
$$-h \leq x_3 \leq 0,$$

and that water occupies the pore space in the subregion defined by the inequalities

$$-h \leq x_3 \leq -h + H(x_1, t).$$

Here, H denotes the height of the water-saturated region, with $0 \leq H(x_1, t) \leq h$. In the region $-h + H(x, t) < x_3 \leq 0$, both air and water occupy the pore space.

To develop a simple, one-dimensional model, assume that the aquifer properties, including $H(x_1, t)$, do not depend on x_2 and that the porosity ϕ and true density γ of the water are constant. With these assumptions, we recast the problem in terms of the spatial variables $x = x_1, y = x_2$, and depth $Z = -x_3$.

We call the locus $Z = h - H(x, t)$ the **water table**. It is a **free surface**, moving vertically in response to water movements. Physically, H plays a role similar to the piezometric head in a confined aquifer. However, as shown in this section, mathematically the water table introduces new difficulties, most notably the fact that it leads to a nonlinear relative of the heat equation. The nonlinearity leads to solutions that differ qualitatively from the Theis solution examined in Section 4.2.

Strictly speaking, the conceptual model underlying Figure 4.8 is too simple. An unconfined aquifer is a mixture consisting of three phases, namely soil, water, and air. In an upper zone, called the **vadose zone**, mobile air, at atmospheric pressure, occupies at least some of the pore space. The water table is the surface that divides the vadose zone from the **saturated zone**, in which only water is mobile. Immediately above the water table is a **capillary fringe**, in which water occupies the pore space at pressures less than atmospheric. Section 6.2 discusses these physics in more detail.

The mechanics of the vadose zone involve nonlinear, multifluid flows. By tracking only the free surface $Z = h - H(x, t)$, we simplify the physics in a way that allows us to solve a single-fluid—but still nonlinear—model. Even with this simplification, the free-surface model is too complex to admit a simple solution. To make the model tractable, in 1857, Jules Dupuit [49] adopted the further approximation that the water velocity in the zone $h - H(x, t) \leq Z \leq h$ is purely horizontal:

$$\mathbf{v} = (v(x, Z, t), 0, 0).$$

This **Dupuit assumption** eliminates the need to analyze the complicated effects of free-surface geometry on the water velocity near the water table.

4.3.1 Derivation of the Boussinesq Equation

Consider the mass balance for water in a time-independent (x, y, Z)-region

$$\mathcal{R} = [x, x + \Delta] \times [0, w] \times [0, h]$$

of the aquifer, assuming that the function $H(x, t)$ defining the water table is continuously differentiable. The accumulation of water in $\mathcal{R}$ equals the net flux of water across the boundary, in accordance with the multiphase extension of Eq. (2.4):

$$\frac{d}{dt} \int_{\mathcal{R}} \phi\gamma \, dv = -\int_{\partial\mathcal{R}} \phi\gamma\mathbf{v} \cdot \mathbf{n} \, da. \tag{4.47}$$

The left side of Eq. (4.47) is the time derivative of a volume integral giving the total mass of water in $\mathcal{R}$. On the right side, $\mathbf{n}$ denotes the unit-length vector orthogonal to the bounding surface $\partial\mathcal{R}$ and pointing outward, and da stands for the element of surface integration.

The left side of Eq. (4.47) reduces as follows:

$$\phi\gamma \frac{d}{dt} \int_x^{x+\Delta} \int_0^w \int_0^h dZ \, dy \, dx = \phi\gamma w \int_x^{x+\Delta} \frac{d}{dt} \int_{h-H(x,t)}^h dZ \, dx. \tag{4.48}$$

Exercise 4.11 *Show that*

$$\frac{d}{dt} \int_{h-H(x,t)}^h dZ = \frac{\partial H}{\partial t}(x, t).$$

By the result of Exercise 4.11, the volume integral in Eq. (4.48) reduces even further:

$$\phi\gamma \frac{d}{dt} \int_x^{x+\Delta} \int_0^w \int_0^h dZ \, dy \, dx = \phi\gamma w \int_x^{x+\Delta} \frac{\partial H}{\partial t}(x, t) \, dx = \phi\gamma w\Delta \frac{\partial H}{\partial t}(\xi, t),$$

for some point $\xi \in [x, x + \Delta]$, by the mean value theorem for integrals.

For the surface integral on the right side of Eq. (4.47), we have

$$\int_{\partial\Omega} \phi\gamma\mathbf{v} \cdot \mathbf{n} \, da = \phi\gamma w \left[\int_{h-H(x+\Delta,t)}^{h} v(x + \Delta, t) dZ - \int_{h-H(x,t)}^{h} v(x, t) dZ \right].$$

Combining these results and dividing by $\phi\gamma w\Delta$ yields

$$\frac{\partial H}{\partial t}(\xi, t) = -\frac{1}{\Delta} \left[\int_{h-H(x+\Delta,t)}^{h} v(x + \Delta, t) \, dZ - \int_{h-H(x,t)}^{h} v(x, t) \, dZ \right].$$

In the limit as $\Delta \to 0$, $\xi \to x$, and we obtain

$$\frac{\partial H}{\partial t}(x, t) = -\frac{\partial}{\partial x} \int_{h-H(x,t)}^{h} v(x, t) \, dZ. \tag{4.49}$$

Exercise 4.12 *Analyze the right side of Eq. (4.49) to derive the following differential mass balance:*

$$\frac{\partial H}{\partial t}(x, t) = -\frac{\partial}{\partial x} [H(x, t)v(x, t)]. \tag{4.50}$$

The next step is to rewrite the velocity using Darcy's law, assuming that ϕ, γ, and the permeability k are constant. Set the pressure at the water table equal to the gauge pressure of the atmosphere, and assume that the water pressure below the water table is hydrostatic:

$$p(x, Z, t) = \gamma g[Z - h + H(x, t)] \quad \text{for } Z \geq h - H(x, t).$$

By Darcy's law,

$$\phi v(x, t) = -\frac{k}{\mu} \frac{\partial}{\partial x} [p(x, Z, t) - \gamma gZ] = -K \frac{\partial H}{\partial x}(x, t),$$

where $K = k\gamma g/\mu$. Substituting this expression for the filtration velocity ϕv into the mass balance equation (4.50) yields the flow equation

$$\boxed{\frac{\partial H}{\partial t} - \frac{K}{\phi} \frac{\partial}{\partial x} \left(H \frac{\partial H}{\partial x} \right) = 0.} \tag{4.51}$$

Equation (4.51) is the **Boussinesq equation**, developed by French mathematician Joseph Valentin Boussinesq [25, 26]. Unlike equations derived in Sections 4.1 and 4.2 of this chapter, the Boussinesq equation is nonlinear. Recognizing this fact, American hydrologist George M. Hornberger et al. [74] were among the first to develop numerical solution methods for the equation. Later in this section, we show that the Boussinesq equation possesses a symmetry that admits self-similar solutions for certain initial conditions, as introduced in Section 4.2.

Exercise 4.13   *Use the dimensionless variables*

$$\xi = \frac{x}{h}, \quad \tau = \frac{Kt}{2\phi h}, \quad u = \frac{H}{h}$$

to reduce Eq. (4.51) *to the nondimensional form*

$$\frac{\partial u}{\partial \tau} - \frac{\partial}{\partial \xi}\left(2u\frac{\partial u}{\partial \xi}\right) = 0. \tag{4.52}$$

4.3.2   The Porous Medium Equation

A close mathematical relative of the Boussinesq equation, the **porous medium equation**, arises from different physics, also involving single-fluid flows in porous media. As a simple class of degenerate nonlinear diffusion equations, the porous medium equation has received close scrutiny from mathematicians. In 2007, Spanish mathematician J.L. Vásquez [151] published an encyclopedic study of this equation.

The porous medium equation is

$$\boxed{\frac{\partial u}{\partial t} - \nabla^2 u^{n+1} = 0,} \tag{4.53}$$

where $n \geq 1, u(\mathbf{x}, t)$ is the unknown function, and ∇^2 signifies the three-dimensional Laplace operator. The relationship between Eq. (4.53) and the Boussinesq equation becomes clear if we use the chain rule to rewrite Eq. (4.53) as follows:

$$\frac{\partial u}{\partial t} - \nabla \cdot [(n+1)u^n \nabla u] = 0.$$

Here, the factor $(n + 1)u^n$ plays the role of a nonlinear diffusion coefficient. The parabolic nature of the PDE degenerates at any point where this coefficient vanishes, that is, where $u(\mathbf{x}, t) = 0$. The nondimensional version (4.52) of the Boussinesq equation, derived in Exercise 4.13, arises when we restrict attention to one space dimension and focus on the special case $n = 1$. However, as the following exercise shows, more general values of n arise in related applications.

Exercise 4.14   *The porous-medium equation, with values of $n \geq 1$, models the density of certain gases flowing through porous media. Derive Eq. (4.53) for the fluid density γ in a single-fluid flow through a porous medium when gravitational effects are negligible, using the following three principles:*

$$\phi\frac{\partial \gamma}{\partial t} + \nabla \cdot (\gamma \phi \mathbf{v}) = 0, \qquad \text{Mass balance,} \tag{4.54}$$

$$\phi \mathbf{v} = -\frac{k}{\mu}\nabla p, \qquad \text{Darcy's law,} \tag{4.55}$$

$$p = p_0 \gamma^n \qquad \text{Equation of state.} \tag{4.56}$$

Equation (4.56) is the equation of state for a gas with reference pressure p_0 and **polytropic index** $n \geq 1$. *Hint: Obtaining the porous medium equation from Eqs. (4.54)–(4.56) requires rescaling time using the change of variables $\tau = \kappa t$, where*

$$\kappa = \frac{kp_0 n}{\mu \phi(n+1)}.$$

4.3.3 A Model Problem with a Self-similar Solution

We now examine a classic initial-value problem involving the nonlinear porous medium equation with $n = 1$:

$$\frac{\partial u}{\partial t} - \nabla \cdot (2u\nabla u) = 0.$$

The problem illustrates a striking departure from the smoothing effects associated with the linear heat equation.

Assume that there is an initial mound of fluid concentrated at the origin, $(x_1, x_2) = (0, 0)$, and that the problem is symmetric about the x_3-axis. Thus $u = u(r, t)$, where r denotes the radial coordinate in cylindrical coordinates, as defined in Appendix B. The PDE thus reduces to

$$\frac{\partial u}{\partial t} - \frac{2}{r}\frac{\partial}{\partial r}\left(ru\frac{\partial u}{\partial r}\right) = 0, \tag{4.57}$$

or, equivalently,

$$\frac{\partial u}{\partial t} - \frac{1}{r}\frac{\partial}{\partial r}\left(r\frac{\partial u^2}{\partial r}\right) = 0. \tag{4.58}$$

Russian mathematician Pelageya Polubarinova-Kochina examined similar problems in a 1952 treatise, later translated into English by Belgian-American engineer Roger De Wiest [123, Chapter VIII]. Polubarinova-Kochina died in 1999 at age 100, shortly after publishing her last scientific paper.

We adopt the following initial and boundary conditions:

$$u(r, 0) = 0, \quad r > 0,$$

$$\lim_{r \to \infty} u(r, t) = 0, \quad t \geq 0.$$

In addition, to model the initial mound, we impose the following integral condition:

$$\int_0^\infty u(r, t)\, r\, dr = 1, \quad t \geq 0. \tag{4.59}$$

(See Eq. (B.4) for integration in polar coordinates.) At $t = 0$, this condition is equivalent to an instantaneous, unit-strength point source—that is, a Dirac δ distribution—concentrated at $(x_1, x_2) = (0, 0)$. Conservation of mass implies that Eq. (4.59) remains true for $t > 0$. Since mass is nonnegative, we seek a solution for which $u(r, t) \geq 0$.

The method of self-similar solutions, introduced in Section 4.2, reveals a symmetry of the PDE (4.57) and leads to a solution to the problem. As shown below, the solution itself possesses a nondifferentiable front that propagates with finite speed—a phenomenon not seen in solutions to the linear heat equation. We encounter sharp, finite-speed fronts again in Chapter 6, in analyzing multiphase flows in porous media.

As in Section 4.2, the symmetry of interest is a group of stretching transformations:

$$\xi = \varepsilon r, \quad \tau = \varepsilon^a t, \quad \eta(\xi, \tau) = \varepsilon^b u(r, t), \tag{4.60}$$

where ε is an arbitrary positive scaling parameter. Such a symmetry exists if there are real exponents a and b for which the following invariance condition holds:

$$\frac{\partial \eta}{\partial \tau} - \frac{2}{\xi}\frac{\partial}{\partial \xi}\left(\xi \eta \frac{\partial \eta}{\partial \xi}\right) = C \times \left[\frac{\partial u}{\partial t} - \frac{2}{r}\frac{\partial}{\partial r}\left(ru\frac{\partial u}{\partial r}\right)\right], \tag{4.61}$$

for some nonzero constant C. In this case, as seen in Section 4.2.4, solutions have the self-similar form

$$u(r, t) = t^{b/a} U(\zeta),$$

for the similarity variable $\zeta = rt^{-1/a}$, and we can find the function U by solving an ordinary differential equation.

We first determine the conditions under which a symmetry of the form (4.60) exists.

Exercise 4.15 *Use the chain rule to show that, under the stretching transformation (4.60),*

$$\frac{\partial \eta}{\partial \tau} = \varepsilon^{b-a}\frac{\partial u}{\partial t}, \quad \frac{\partial \eta}{\partial \xi} = \varepsilon^{b-1}\frac{\partial u}{\partial x},$$

and hence

$$\frac{\partial u}{\partial t} - \frac{2}{r}\frac{\partial}{\partial r}\left(ru\frac{\partial u}{\partial r}\right) = \varepsilon^{a-b}\frac{\partial \eta}{\partial \tau} - \varepsilon^{2-2b}\frac{2}{\xi}\frac{\partial}{\partial \xi}\left(\xi \eta \frac{\partial \eta}{\partial \xi}\right).$$

The invariance condition (4.61) therefore requires that $\varepsilon^{a-b} = \varepsilon^{2-2b}$ for all $\varepsilon > 0$. It follows that $a = 2 - b$. In this case, we seek a self-similar solution of the form

$$u(r, t) = t^{b/(2-b)} U(\zeta), \quad \zeta = rt^{-1/(2-b)}.$$

To determine what values b, and hence a, can take, impose the mass conservation condition (4.59):

$$1 = \int_0^\infty u(r, t)\, r\, dr = t^{b/(2-b)} \int_0^\infty U\left(rt^{-1/(2-b)}\right) r\, dr$$

$$= t^{b/(2-b)} t^{2/(2-b)} \int_0^\infty U(y) \, y \, dy,$$

the last identity following from the change of variables $y = rt^{-1/(2-b)}$. This equation requires the right side to be independent of t, which implies that $b = -2$. It follows that $a = 4$, and the similarity variable for the axisymmetric porous medium equation (4.58) with $n = 1$ is $\zeta = rt^{-1/4}$.

Next we determine the ordinary differential equation that governs $U(\zeta)$.

Exercise 4.16 *Substitute the newly found form $u(r,t) = t^{-1/2} U(rt^{-1/4})$ into Eq. (4.58) and simplify to obtain*

$$(U^2)'' + \frac{1}{\zeta}(U^2)' + \frac{1}{4}U' + \frac{1}{2}U = 0. \tag{4.62}$$

Then multiply Eq. (4.62) by ζ to get the simpler equation

$$\left[\zeta(U^2)' \right]' + \frac{1}{4}(\zeta^2 U)' = 0.$$

Integrating this differential equation yields

$$\zeta(U^2)' + \frac{1}{4}\zeta^2 U = C_1. \tag{4.63}$$

To determine the constant C_1 of integration, observe that the left side of Eq. (4.63) tends to 0 as $\zeta \to 0$, so $C_1 = 0$.

Exercise 4.17 *Differentiate $(U^2)'$ in Eq. (4.63), then conclude that the equation holds if $U = 0$ or $U' = -\zeta/8$, that is, $U = -\frac{1}{16}\zeta^2 + C_2$.*

To construct a solution that satisfies the condition $u(x,t) \geq 0$, we follow Barenblatt [12] and define $U(\zeta)$ piecewise:

$$U(\zeta) = \begin{cases} \dfrac{1}{16}\left(\zeta_0^2 - \zeta^2\right), & \text{if } -\zeta_0 \leq \zeta \leq \zeta_0; \\ 0, & \text{otherwise.} \end{cases}$$

Mathematically, $\zeta_0^2/16$ is the constant C_2 of integration. Physically, ζ_0 represents a point on the ζ-axis beyond which the solution vanishes, that is, a front. To determine ζ_0, we appeal once again to Eq. (4.59).

Exercise 4.18 *Impose the mass conservation condition*

$$\int_0^{\zeta_0} \frac{1}{16}\left(\zeta_0^2 - \zeta^2\right) \zeta \, d\zeta = 1$$

to show that $\zeta_0 = 2^{3/2}$. Hence,

$$U(\zeta) = \begin{cases} \dfrac{1}{16}\left(8 - \zeta^2\right), & \text{if } 0 \le \zeta \le 2^{3/2}, \\ 0, & \text{otherwise.} \end{cases}$$

This expression yields the following solution: for $t > 0$,

$$u(r, t) = \begin{cases} \dfrac{1}{16}t^{-1/2}\left(8 - r^2 t^{-1/2}\right), & \text{if } 0 \le r \le 2^{3/2}t^{1/4}; \\ 0, & \text{otherwise.} \end{cases} \tag{4.64}$$

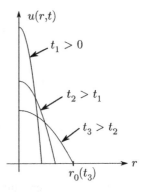

Figure 4.9 shows graphs of this solution at several times. Because the solution's derivative $\partial u / \partial r$ is discontinuous at the locus $r_0(t) = 2^{3/2}t^{1/4}$ of the advancing front, $u(r, t)$ differs qualitatively from the infinitely differentiable classical solutions that we expect for the linear heat equation. In fact, strictly speaking, the discontinuity in $\partial u / \partial r$ implies that Eq. (4.64) cannot be a solution to the PDE (4.57), which requires $u(r, t)$ to be twice differentiable with respect to r.

To resolve this difficulty, we must admit **weak solutions**. These do not necessarily satisfy the PDE in the literal or **classical** sense; instead, they allow for discontinuities in the function or its derivatives. Section 5.2 defines weak solutions more precisely.

Also contrasting with solutions to the heat equation is the finite propagation speed $r_0'(t) = t^{-3/4}/\sqrt{2}$ of the advancing front. Figure 4.10 shows r_0 as a function of t. At any instant in time, the initial condition has had no influence beyond this front.

Figure 4.9 Weak solution to the porous medium equation showing a nondifferentiable front $\pm r_0(t)$ propagating outward from the origin.

As this example illustrates, nonlinear diffusion equations with degenerate diffusion coefficients can possess solutions that differ significantly and qualitatively from solutions to the linear heat equation, for which solutions are infinitely smooth and propagate with infinite speed. Solutions with moving fronts and finite propagation speed arise much more generally in solutions to the porous medium equation and to other nonlinear extensions of the heat equation. Barenblatt et al. [13, Section 3.4.3] develop exact solutions of this type for cases in which the exponent $n > 1$ and in rectilinear, axisymmetric, and spherically symmetric geometries. Section 6.2 revisits these phenomena, in connection with variably saturated flows in porous media.

Figure 4.10 Location of the advancing front $r_0(t) = 2^{3/2}t^{1/4}$ as a function of time.

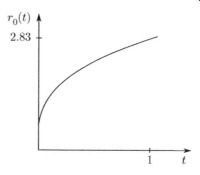

Exercise 4.19 *Consider the porous medium equation (4.53) in three dimensions. In a problem having radial symmetry—for example a problem with a point source at the origin—the solution has the form $u(r, t)$, where $r = \|\mathbf{x}\|$ denotes distance from the origin. In this case, a transformation to spherical coordinates (see Exercise 111) converts the PDE to*

$$\frac{\partial u}{\partial t} - \frac{1}{r^2}\frac{\partial}{\partial r}\left[r^2\frac{\partial}{\partial r}\left(u^{n+1}\right)\right] = 0,$$

where $n \geq 1$.

(A) *Reduce this PDE to the equivalent form*

$$\frac{\partial u}{\partial t} - \frac{2}{r}\frac{\partial}{\partial r}\left(u^{n+1}\right) - \frac{\partial^2}{\partial r^2}\left(u^{n+1}\right) = 0.$$

(B) *Find a group of stretching transformations (4.60) that leaves this equation invariant.*

(C) *Conclude that $u(r, t) = t^{1/(n-3)}U(\zeta)$, where $\zeta = rt^{1/(n-3)}$.*

5

Solute Transport

5.1 The Transport Equation

Many underground flow applications require tracking the movements of particular constituents in one or more fluid phases. A water-soluble contaminant undergoing transport in a groundwater aquifer furnishes a prototypical example. In many enhanced oil recovery processes, injected fluids transport constituents that alter the chemistry of the fluid–rock system, promoting more efficient oil recovery.

To model these effects, we must take into account a set of miscible species, indexed as $i = 1, 2, \ldots, N$, distributed among two or more immiscible phases. For example, consider the case of a soluble groundwater contaminant that adsorbs onto the aquifer rock. We treat the fluid phase $\alpha = F$ and the solid phase $\alpha = R$ as distinct phases and identify three species:

- Rock ($i = 1$)
- Water ($i = 2$)
- Contaminant ($i = 3$).

In this framework, each ordered pair (i, α) of indices is a constituent. We must therefore track the motions of six constituents:

$$(1, F), \ (2, F), \ (3, F) \quad \text{and} \quad (1, R), \ (2, R), \ (3, R).$$

Throughout this chapter, we assume that constituents $(1, F)$ and $(2, R)$ are absent—that is, that the rock does not dissolve into the fluid phase and that there is no water in the solid phase. The general framework allows one to relax these assumptions, for example to accommodate the transport, sublimation, and freezing of water vapor in snow.

The Mathematics of Fluid Flow Through Porous Media, First Edition. Myron B. Allen.
© 2021 John Wiley & Sons, Inc. Published 2021 by John Wiley & Sons, Inc.

5.1.1 Mass Balance of Miscible Species

For the mass balance, Eqs. (2.29) and (2.30) yield

$$\frac{\partial}{\partial t}(\phi_\alpha \gamma_{(i,\alpha)}) + \nabla \cdot (\phi_\alpha \gamma_{(i,\alpha)} \mathbf{v}_{(i,\alpha)}) = r_{(i,\alpha)}, \tag{5.1}$$

for each constituent (i, α), subject to the restriction

$$\sum_{i=1}^{N} \sum_{\alpha} r_{(i,\alpha)} = 0.$$

In Eq. (5.1), $\gamma_{(i,\alpha)}$ denotes the true density of constituent (i, α), and each term has dimension $ML^{-3}T^{-1}$. It is common to rewrite this mass balance equation in terms of dimensionless mass fractions, defined as

$$\omega_{(i,\alpha)} = \gamma_{(i,\alpha)}/\gamma_\alpha,$$

where γ_α denotes the true density of phase α.

Exercise 5.1   *Justify the following identities:*

$$\sum_{i=1}^{N} \omega_{(i,\alpha)} = 1, \quad \sum_{i-1}^{N} \omega_{(i,\alpha)} \mathbf{v}_{(i,\alpha)} = \mathbf{v}_\alpha,$$

where $\mathbf{v}_\alpha$ is the spatial velocity of phase α.

In terms of mass fractions, the mass balance (5.1) becomes

$$\frac{\partial}{\partial t}(\phi_\alpha \gamma_\alpha \omega_{(i,\alpha)}) + \nabla \cdot (\phi_\alpha \gamma_\alpha \omega_{(i,\alpha)} \mathbf{v}_{(i,\alpha)}) = r_{(i,\alpha)}. \tag{5.2}$$

Summing this equation over the species index i yields the overall mass balance for phase α:

$$\frac{\partial}{\partial t}(\phi_\alpha \gamma_\alpha) + \nabla \cdot (\phi_\alpha \gamma_\alpha \mathbf{v}_\alpha) = r_\alpha = \sum_{i=1}^{N} r_{(i,\alpha)},$$

where the sum on the right represents the net exchange of mass from all species into phase α from other phases.

Two additional definitions convert Eq. (5.2) to a form that commonly appears in applications. Define the **concentration** of constituent (i, α) as

$$c_{i,\alpha} = \gamma_\alpha \omega_{(i,\alpha)}, \tag{5.3}$$

having dimension (mass of (i, α))/(volume of α) or ML^{-3}, and the **diffusion velocity** of species i in phase α as

$$v_{(i,\alpha)} = \mathbf{v}_{(i,\alpha)} - \mathbf{v}_\alpha, \tag{5.4}$$

by analogy with the diffusion velocity defined in Eq. (2.27). In terms of these quantities, the constituent mass balance has the following form:

$$\underbrace{\frac{\partial}{\partial t}(\phi_\alpha c_{(i,\alpha)})}_{(I)} + \underbrace{\nabla \cdot (\phi_\alpha \mathbf{v}_\alpha c_{(i,\alpha)})}_{(II)} + \underbrace{\nabla \cdot \mathbf{j}_{(i,\alpha)}}_{(III)} = \underbrace{r_{(i,\alpha)}}_{(IV)} .$$

(5.5)

Here,

$$\mathbf{j}_{(i,\alpha)} = \phi_\alpha c_{(i,\alpha)} \mathbf{v}_{(i,\alpha)}$$

(5.6)

is the **diffusive flux** of species i in phase α, having dimension $ML^{-2}T^{-1}$. As in Section 2.5, we refer to the terms labeled (I), (II), (III), and (IV) as the **accumulation**, **advection**, **diffusion**, and **reaction** terms, respectively.

The remainder of this chapter focuses on the special case involving the transport of a single species i in a single fluid F flowing through a porous medium composed of a rock phase R. This restriction allows us to simplify the notation by stripping away subscripts, writing Eq. (5.5) as follows:

$$\frac{\partial}{\partial t}(\phi c) + \nabla \cdot (\phi \mathbf{v} c) + \nabla \cdot \mathbf{j} = r.$$

(5.7)

Here, $\phi = \phi_F$ is the porosity, $c = c_{(i,F)}$ is the concentration of species i in the fluid, $\mathbf{v} = \mathbf{v}_F$ denotes the fluid velocity, $\mathbf{j} = \mathbf{j}_{(i,F)}$, and $r = r_{(i,F)}$. Chapter 7 examines more complicated flows, in which several species undergo mass transfer among several fluid phases.

5.1.2 Hydrodynamic Dispersion

Mysteries lurk in the diffusion term for porous-medium flows. In 1961, Swiss geophysicist Adrian Scheidegger [133] and Israeli hydrologist Jacob Bear [17] proposed a tensor model to account for the effects of microscopic pore geometry in the spreading of solutes. French hydrologists Jean J. Fried and M.A. Combarnous [55] provide a detailed review of this classical theory, referred to as **hydrodynamic dispersion**.

The theory, which remains in common use, begins with the decomposition

$$\mathbf{j} = \mathbf{j}_{mol} + \mathbf{j}_{mech},$$

where $\mathbf{j}_{mol}$ accounts for the molecular diffusion of the solute and $\mathbf{j}_{mech}$ accounts for the mechanical mixing effects associated with the microscopic geometry of the porous medium. Molecular diffusion is the most straightforward of these effects to model: We adopt an equation developed by German physician Adolf Fick [52]. **Fick's law** has the form

$$\mathbf{j}_{mol} = -\phi D_M \nabla c,$$

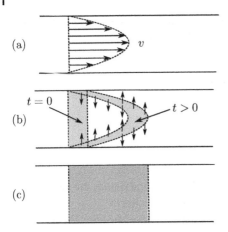

Figure 5.1 Qualitative illustration of Taylor diffusion. (a) Parabolic velocity profile given by the Hagen–Poiseuille solution for flow in a thin cylindrical tube. (b) Transport of an initial solute slug under the influence of pure advection, with arrows showing the directions of diffusive transport by molecular diffusion when $t > 0$. (c) Slug at $t > 0$ resulting from radially varying longitudinal advection coupled with transverse diffusion.

where D_M denotes the **molecular diffusion coefficient**, a positive parameter having dimension L^2T^{-1}. The factor ϕ models the interference of the rock matrix, which retards the spread of solute in the fluid.

The mechanical mixing effects modeled by the term $\mathbf{j}_{mech}$ are more complicated [64]. Perhaps the most rigorously analyzed of these effects is **Taylor diffusion**. This phenomenon, explained by British fluid mechanician Sir G.I. Taylor [144], results in longitudinal spreading of solute in a thin tube, as a result of radial solute diffusion driven by the parabolic Hagen–Poiseuille velocity profiles analyzed in Section 2.4.

Figure 5.1 illustrates the effect qualitatively. In the absence of molecular diffusion, an initial slug having uniform concentration would advect longitudinally, yielding the parabolic concentration profile shown in Figure 5.1b. But diffusion causes mass from the solute to spread in directions transverse to the tube's axis, yielding a slug of nonzero solute concentration that is longer in the axial direction than the original slug. In short, longitudinal advection, varying according to the Hagen–Poiseuille solution (2.21), couples with transverse molecular diffusion governed by Fick's law to yield a longitudinal spreading effect.

Other effects that contribute to mechanical mixing include variable path lengths, which result in different longitudinal transit times for fluid particles that start at nearby positions, and stream splitting, in which initially nearby fluid particles follow different pore channels through the rock. Figure 5.2 shows these effects schematically.

The Bear–Scheidegger model of these effects rests on four premises, based on the conceptual decomposition

$$\mathbf{j}_{mech} = \mathbf{j}_{long} + \mathbf{j}_{tran}.$$

Figure 5.2 Mechanical dispersion effects. (a) Solute spreading as a result of variable path lengths. (b) Solute spreading as a result of stream splitting.

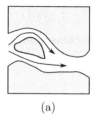

(a) (b)

Here, $\mathbf{j}_{\text{long}}$ captures the solute spreading longitudinally (parallel to the filtration velocity $\phi\mathbf{v}$), and $\mathbf{j}_{\text{tran}}$ models transverse spreading.

1. The magnitude of the longitudinal flux increases linearly with the filtration velocity:

$$\|\mathbf{j}_{\text{long}}\| = \alpha_L \|\phi\mathbf{v}\| \|\nabla c\| . \tag{5.8}$$

Here, the **longitudinal dispersivity** α_L is a positive parameter having dimension L.

2. Similarly,

$$\|\mathbf{j}_{\text{tran}}\| = \alpha_T \|\phi\mathbf{v}\| \|\nabla c\|, \tag{5.9}$$

where the **transverse dispersivity** α_T is a positive parameter also having dimension L.

3. The molecular diffusion coefficient D_M is much smaller in magnitude than $\alpha_T \|\phi\mathbf{v}\|$, and $\alpha_T < \alpha_L$.

4. The dispersivities α_L, α_T are properties of the rock matrix that are independent of the mechanical properties of the fluid phase.

The fourth premise is problematic. Empirical studies indicate that the values of α_L and α_T depend not only on the rock but also on the spatial scale of observation, increasing in magnitude as the scale increases. At the laboratory bench scale—roughly 1 m—dispersivities in the range 10^{-2}–10 m are typical; at the field scale—say 10^2 m—dispersivities may range from 10 to 10^2 m [59,108]. This scale dependence arises from spatial heterogeneity in such properties as permeability or hydraulic conductivity [58, Chapter 5]. Owing to uncertainty about subsurface media, we can quantify these heterogeneities and the resulting fluid-velocity fluctuations only through statistics such as means, covariances, and correlation lengths. The parameters α_L and α_T, when modeled by Eqs. (5.8) and (5.9) and measured at the bench scale, cannot account for the velocity fluctuations that occur because of heterogeneities at larger scales.

This conundrum notwithstanding, the following standard model for hydrodynamic dispersion incorporates the observations listed above in a tensor form:

$$\mathbf{j} = -\phi\mathbf{D}\nabla c,$$

$$D = D_M I + \alpha_L \|\mathbf{v}\| W + \alpha_T \|\mathbf{v}\| (I - W). \tag{5.10}$$

Here, I denotes the identity tensor, and

$$W = \frac{1}{\|\mathbf{v}\|^2} \mathbf{v} \otimes \mathbf{v}.$$

Exercise 5.2 *Use the definition (2.33) of dyadic products and the result of Exercise 2.17 to show that, with respect to any orthonormal basis, the tensor W has matrix representation*

$$\frac{1}{\|\mathbf{v}\|^2} \begin{bmatrix} v_1 v_1 & v_1 v_2 & v_1 v_3 \\ v_2 v_1 & v_2 v_2 & v_2 v_3 \\ v_3 v_1 & v_3 v_2 & v_3 v_3 \end{bmatrix}.$$

With this model of hydrodynamic dispersion, Eq. (5.7) becomes the **advection–diffusion-reaction equation**:

$$\frac{\partial}{\partial t}(\phi c) + \nabla \cdot (\phi \mathbf{v} c) - \nabla \cdot (\phi D \nabla c) = r. \tag{5.11}$$

In many applications, there is a fixed value $\bar{c}$ that characterizes the transport, such as the concentration at a source or the maximum observed concentration. In these cases, it is common to replace the concentration $c(\mathbf{x}, t)$ by a dimensionless **normalized concentration** $c(\mathbf{x}, t)/\bar{c}$. This numerically convenient substitution leaves Eq. (5.11) unchanged.

As remarks above suggest, the scale dependence of the dispersivities α_L and α_T has led to skepticism about whether a model as simple as Eq. (5.10) can adequately model the spreading of solutes in underground fluid flows. American engineer Lynn Gelhar et al. [59] and Israeli hydrologist Gedeon Dagan [41,42] have pioneered ongoing, rigorous efforts to understand how diffusive fluxes depend on the spatial length scales of heterogeneity in porous media. Consequently, even though Eq. (5.11) remains in wide use, many open questions persist about the best way to characterize hydrodynamic dispersion mathematically.

5.2 One-Dimensional Advection

To understand the properties of solutions to Eq. (5.11), it helps to examine simple cases in which $r = 0$, ϕ is constant, and $\nabla \cdot \mathbf{v} = 0$. These hypotheses reduce Eq. (5.11) to the **advection–diffusion equation**:

$$\boxed{\frac{\partial c}{\partial t} + \mathbf{v} \cdot \nabla c - \nabla \cdot (D \nabla c) = 0.} \tag{5.12}$$

Further restricting attention to one space dimension in which v and D are positive constants reduces Eq. (5.12) to

$$\frac{\partial c}{\partial t} + v\frac{\partial c}{\partial x} - D\frac{\partial^2 c}{\partial x^2} = 0. \tag{5.13}$$

The inequality $v > 0$ implies that fluid flows from left to right.

This linear partial differential equation (PDE) is parabolic in the concentration (or normalized concentration) c, so the appropriate auxiliary conditions are the same types as those reviewed in Section 3.4 for the time-dependent groundwater flow equation. For the same reason, we expect diffusion to exert a smoothing effect on the concentration profile $c(x, t)$ as t increases, by analogy with the parabolic heat equation discussed in Section 4.2.

However, in practice, owing to competition between the advection and diffusion terms, Eq. (5.13) exhibits a split personality. To weigh the two effects in a fashion that is independent of the subjective choice of physical units, we follow an approach introduced in Section 2.3, converting the equation to dimensionless form. Consider an application in which there exists a characteristic length L, such as the length of a flow channel or the diameter of an inlet.

Exercise 5.3 *Define dimensionless space and time variables*

$$\xi = \frac{x}{L}, \quad \tau = \frac{vt}{L}.$$

Show that Eq. (5.13) is equivalent to the following PDE:

$$\frac{\partial c}{\partial \tau} + \frac{\partial c}{\partial \xi} - \frac{1}{\text{Pe}}\frac{\partial^2 c}{\partial \xi^2} = 0. \tag{5.14}$$

Here, $\text{Pe} = vL/D$ *is the dimensionless **Péclet number**, named for nineteenth-century French physicist Jean Claude Eugène Péclet.*

5.2.1 Pure Advection and the Method of Characteristics

When Pe is much larger than 1, advection dominates the smoothing effects of diffusion. In cases where diffusion is negligible compared with advection, for example in flows where the filtration velocity is large in response to large applied pressure gradients, the dimensionless transport equation (5.13) reduces to the **advection equation**

$$\frac{\partial c}{\partial \tau} + \frac{\partial c}{\partial \xi} = 0,$$

or, in dimensional form,

$$\boxed{\frac{\partial c}{\partial t} + v\frac{\partial c}{\partial x} = 0.} \tag{5.15}$$

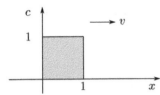

Figure 5.3 Initial concentration profile $c(x, 0)$ for a sample initial-value problem involving pure advection.

This limiting case is a first-order hyperbolic PDE, whose solutions differ qualitatively from those arising when the diffusion term is present.

To see the difference, consider an initial-value problem for Eq. (5.15), posed for $-\infty < x < \infty$ and $t > 0$ and having initial condition

$$c(x, 0) = c_0(x) = \begin{cases} 1, & \text{if } 0 \leq x \leq 1, \\ 0, & \text{otherwise.} \end{cases} \tag{5.16}$$

This problem models diffusion-free solute transport along an infinite aquifer, with the solute initially confined to a spatially uniform slug between $x = 0$ and $x = 1$, as shown in Figure 5.3. We have posed the condition (5.16) along an **initial curve** in the (t, x)-plane, namely the curve $t = 0$. Section 5.2.2 discusses appropriate auxiliary conditions for Eq. (5.15) more generally.

Wherever the concentration $c(x, t)$ is sufficiently smooth, it defines a surface $(x, t, c(x, t))$ over the (x, t)-plane, as sketched in Figure 5.4. Consider any continuously differentiable path $(x(s), t(s))$ in the (x, t)-plane for which the derivative $(x'(s), t'(s))$ never vanishes. Wherever $c(x, t)$ is differentiable, the chain rule gives its rate of change along such a path as

$$t'(s)\frac{\partial c}{\partial t} + x'(s)\frac{\partial c}{\partial x} = \frac{dc}{ds}. \tag{5.17}$$

The left side of Eq. (5.17) is the **directional derivative** of $c(x, t)$ with respect to the vector $(x'(s), t'(s))$ tangent to the path.

Of special interest are paths $(x(s), t(s))$ along which the left side of the PDE (5.15) coincides with the directional derivative (5.17). Comparing the left sides of Eqs. (5.15) and (5.17) shows that one can construct such paths by imposing the conditions

$$t'(s) = 1, \quad x'(s) = v,$$

or, more simply,

$$\frac{x'(s)}{t'(s)} = \frac{dx}{dt} = v. \tag{5.18}$$

This ordinary differential equation has general solution $x = vt + C$, where C is a constant of integration. According to Eq. (5.17),

$$\frac{dc}{ds} = 0 \tag{5.19}$$

along such paths.

Figure 5.4 Geometry of the method of characteristics, showing a hypothetical solution surface $(x, t, c(x, t))$ and a differentiable path $(x(s), t(s))$ in the (x, t)-plane.

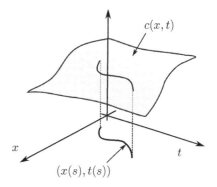

The ordinary differential equation (5.19) is the **characteristic equation** for Eq. (5.15); Eq. (5.18) defines a family of paths called **characteristic curves**. Thus, we have reduced the PDE (5.15) to an ordinary differential equation describing how the concentration changes as we move along a characteristic curve. In particular, c remains constant along paths in the (x, t)-plane for which $x = vt + C$. To determine the concentration value $c(x, t)$, follow the characteristic curve that passes through (x, t) to the point where the curve intersects the initial curve, where a concentration value is prescribed.

Exercise 5.4 *Use this reasoning, called the **method of characteristics**, to show that the function $c(x, t) = c_0(x - vt)$ solves the PDE (5.15), subject to the initial condition (5.16). Sketch the solution at several values of $t > 0$.*

In the solution found in Exercise 5.4, the contaminant slug moves downstream with speed v, its shape remaining unchanged. There is no smoothing; sharp concentration fronts remain intact as t increases.

Exercise 5.5 *Solve the **advection-reaction equation***

$$\frac{\partial c}{\partial t} + v\frac{\partial c}{\partial x} = -\kappa c,$$

subject to the initial condition (5.16). Here κ signifies a constant, positive decay coefficient. In what physical settings might such an equation arise?

5.2.2 Auxiliary Conditions for First-Order PDEs

As with the time-dependent and steady flow equations derived in Chapter 3, certain types of auxiliary conditions yield well posed problems for first-order PDEs, of which the advection equation (5.15) is the simplest example. The method of characteristics holds the key: We can prescribe values of the solution along any curve

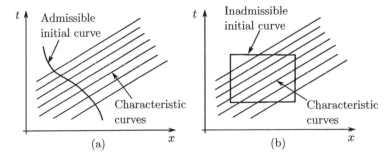

Figure 5.5 (a) An admissible initial curve for a first-order PDE. This curve is nowhere tangent to the characteristic curves, and no characteristic curve intersects it more than once. (b) An initial curve that is inadmissible, because characteristic curves intersect it more than once.

in the (x, t)-plane that is nowhere tangent to a characteristic curve and that no characteristic curve intersects more than once. These conditions ensure that solution values propagating along characteristic curves, according to the characteristic equation, do not conflict with other prescribed initial or boundary values.

From a physical perspective, when $v > 0$, this restriction admits conditions of the following types:

- Initial conditions, such as prescribed values of $c(x, t)$ on some interval of the line $t = 0$;
- Boundary conditions, such as prescribed values of $c(x, t)$ on some interval of the line $x = 0$;
- Some combination of initial and boundary conditions, such as prescribed values of $c(x, t)$ along the union of the nonnegative x- and t-axes.

Mathematicians refer to all such curves as **initial curves** and to the problem consisting of the first-order PDE together with prescribed data along an initial curve as a **Cauchy problem**. From this more mathematical perspective, the range of possible initial curves is quite broad. Figure 5.5a shows an admissible initial curve in the (x, t)-plane—a curve that would be hard to characterize using strictly physical interpretations of the terms boundary condition and initial condition. Figure 5.5b shows a boundary-like initial curve that is inadmissible, because characteristic curves intersect it more than once.

5.2.3 Weak Solutions

The initial condition (5.16) is not differentiable, and as a consequence the method of characteristics yields a solution to the PDE (5.15) that is not differentiable. Such a function cannot be a solution in the classical sense, since it lacks the smoothness required to allow for partial differentiation throughout the (x, t)-domain. Instead, it is a **weak solution**—a term encountered in Section 4.3. To make this term precise, consider the following exercise.

Exercise 5.6 *Let $c(x, t)$ be a solution to the initial-value problem given by Eqs. (5.15) and (5.16). Let $\varphi(x, t)$ be any infinitely differentiable **test function** such that $\varphi \to 0$ as $|x| \to \infty$ and also as $t \to \infty$. Use integration by parts to show that*

$$\int_0^\infty \int_{-\infty}^\infty c \left(\frac{\partial \varphi}{\partial t} + v \frac{\partial \varphi}{\partial x} \right) \, dx \, dt + \int_{-\infty}^\infty c_0(x) \varphi(x, 0) \, dx = 0. \tag{5.20}$$

Thus, multiplication by the smooth test function φ and integration by parts shift differential operators from the solution c to the smooth function φ. Observe that Eq. (5.20) not only holds for classical solutions but can also make sense in cases when c is not differentiable.

A function $c(x, t)$ is a **weak solution** of the initial-value problem if Eq. (5.20) holds for every infinitely differentiable function $\varphi(x, t)$ that vanishes outside some bounded region in the (x, t)-plane. According to this definition, every classical solution is a weak solution. But the weak formulation admits solutions that fail to satisfy some smoothness conditions that, in a strict sense, the PDE requires.

The following exercise shows how to extend the notion of weak solution to second-order PDEs, such as those encountered in Section 4.3.

Exercise 5.7 *Consider the porous medium equation,*

$$\frac{\partial u}{\partial t} - \nabla^2(u^n) = 0,$$

posed for $t > 0$ on a simply connected region $\mathcal{R}$ in three-dimensional space having a smooth boundary $\partial \mathcal{R}$. Impose an initial condition $u(\mathbf{x}, 0) = u_0(\mathbf{x})$. Let $\varphi(\mathbf{x}, t)$ be any infinitely differentiable test function that vanishes on $\partial \mathcal{R}$ and for which $\varphi \to 0$ as $t \to \infty$. Suppose $u(\mathbf{x}, t)$ is a solution.

(A) *Use integration by parts to show that*

$$\int_0^\infty \frac{\partial u}{\partial t} \varphi \, dt = - \int_0^\infty u \frac{\partial \varphi}{\partial t} \, dt - u_0(\mathbf{x}) \varphi(\mathbf{x}, 0).$$

(B) *Recall that $\nabla^2 = \nabla \cdot \nabla$. By the product rule, $\nabla \cdot (f \nabla g) = \nabla f \cdot \nabla g + f \nabla^2 g$ for sufficiently differentiable functions f and g. Prove that*

$$\int_{\mathcal{R}} \varphi \nabla^2 u^n \, dv = \int_{\partial \mathcal{R}} \varphi \nabla u^n \cdot \mathbf{n} \, d\sigma - \int_{\mathcal{R}} \nabla u^n \cdot \nabla \varphi \, dv. \tag{5.21}$$

Here $\mathbf{n}$ denotes the unit-length vector field on $\partial \mathcal{R}$ pointing outward from $\mathcal{R}$. Because the test function φ vanishes on $\partial \mathcal{R}$, the identity (5.21) reduces to

$$\int_{\mathcal{R}} \varphi \nabla^2 u^n \, dv = - \int_{\mathcal{R}} \nabla u^n \cdot \nabla \varphi \, dv.$$

The results of Exercise 5.7 suggest the following definition: A function $u(\mathbf{x}, t)$ is a weak solution to the porous medium equation on $\mathcal{R}$ with initial condition $u(\mathbf{x}, 0) = u_0(\mathbf{x})$ if

$$\int_{\mathcal{R}} \int_0^\infty \left(u \frac{\partial \varphi}{\partial t} - \nabla u^n \cdot \nabla \varphi \right) \, dt \, dv + \int_{\mathcal{R}} u_0(\mathbf{x}) \varphi(\mathbf{x}, 0) \, dv, \tag{5.22}$$

for all infinitely differentiable test functions $\varphi(\mathbf{x}, t)$ that vanish on $\partial \mathcal{R}$ and as $t \to \infty$. Equation (5.22) clearly allows solutions that are not twice differentiable. It also allows solutions in which ∇u^n exhibits discontinuities, since it requires only that the function $\nabla u^n \cdot \nabla \varphi$ be integrable.

We encounter weak solutions again later in this chapter and in Chapter 6.

5.3 The Advection–Diffusion Equation

Next, we examine the effect that a positive diffusion coefficient D has on solutions of the one-dimensional, constant-coefficient advection–diffusion equation (5.13). This equation possesses classical solutions. To begin the derivation, we adopt a moving coordinate system:

$$\xi(x, t) = x - vt, \quad \tau(x, t) = t. \tag{5.23}$$

Exercise 5.8 *In terms of the variables ξ, τ introduced in Eq. (5.23), define $\hat{c}(\xi(x, t), \tau(x, t)) = c(x, t)$. Use the chain rule to show that Eq. (5.14) implies the following one-dimensional heat equation for $\hat{c}$:*

$$\frac{\partial \hat{c}}{\partial \tau} - D \frac{\partial^2 \hat{c}}{\partial \xi^2} = 0. \tag{5.24}$$

We examine two solutions to this equation, each associated with a different initial condition imposed on the infinite spatial domain $-\infty < \xi < \infty$.

5.3.1 The Moving Plume Problem

In the **moving plume** problem, the initial condition is

$$\hat{c}(\xi, 0) = \delta(\xi). \tag{5.25}$$

Here, δ denotes the one-dimensional Dirac δ distribution, whose defining property parallels that of its two-dimensional analog, given in Eq. (4.8): If $\mathcal{R}$ is any interval on the real line and φ is any smooth function, then

$$\int_{\mathcal{R}} \varphi(\xi) \delta(\xi - y) \, d\xi = \begin{cases} \varphi(y), & \text{if } y \in \mathcal{R}, \\ 0, & \text{otherwise.} \end{cases}$$

Physically, the initial condition (5.25) represents an idealized, point-source limit of the initial condition in Eq. (5.16), with all the contaminant mass concentrated at the point $\xi = 0$ at time $\tau = 0$.

The initial condition (5.25) together with the origins of the PDE as a mass balance imply the following mass conservation constraint: For any $\tau > 0$,

$$\int_{-\infty}^{\infty} \hat{c}(\xi, \tau)\, d\xi = \int_{-\infty}^{\infty} \hat{c}(\xi, 0)\, d\xi = \int_{-\infty}^{\infty} \delta(\xi)\, d\xi = 1. \tag{5.26}$$

Also, for any finite value of τ we expect the solution to decay far from the source:

$$\lim_{|\xi| \to \infty} \hat{c}(\xi, \tau) = 0 = \lim_{|\xi| \to \infty} \frac{\partial \hat{c}}{\partial \xi}(\xi, \tau). \tag{5.27}$$

We solve the initial-value problem defined by Eqs. (5.24) and (5.25) using results from the method of self-similar solutions, described in Section 4.2.

Exercise 5.9 *Show that the heat equation (5.24) possesses a symmetry in the form of a stretching transformation and that self-similar solutions to this equation have the form*

$$\hat{c}(\xi, \tau) = \tau^{b/2} U(\zeta), \tag{5.28}$$

for some constant b, where $\zeta = \xi \tau^{-1/2}$ is a similarity variable.

Exercise 5.10 *Substitute the expression (5.28) for $\hat{c}$ into the PDE (5.24) to conclude that $U(\zeta)$ must satisfy the ordinary differential equation*

$$U''(\zeta) + \frac{\zeta}{2D} U'(\zeta) + \frac{b}{2D} U(\zeta) = 0. \tag{5.29}$$

To solve this equation, it helps to determine a value for the exponent b.

Exercise 5.11 *Using the representation (5.28), show that*

$$\int_{-\infty}^{\infty} \hat{c}(\xi, \tau)\, d\xi = \tau^{b/2+1/2} \int_{-\infty}^{\infty} U(y)\, dy. \tag{5.30}$$

Equation (5.30) shows that, for nonzero U, the integral has a constant value, as the condition (5.26) requires, only if $b = -1$.

The ordinary differential equation (5.29) therefore reduces to

$$U'' + \frac{\zeta}{2D} U' + \frac{1}{2D} U = U'' + \frac{1}{2D}(\zeta U)' = 0.$$

It follows that $U' + \zeta U/(2D)$ is constant. The decay conditions (5.27) imply that $U(\zeta) \to 0$ and $U'(\zeta) \to 0$ as $|\zeta| \to \infty$, which is possible only if this constant is 0. Therefore,

$$U' + \frac{1}{2D} \zeta U = 0. \tag{5.31}$$

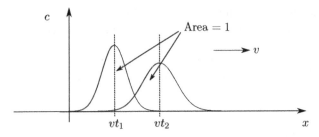

Figure 5.6 The solution $c(x, t)$ to the moving plume problem at two different times.

Exercise 5.12 *Verify that Eq. (5.31) has general solution*

$$U(\zeta) = C \exp\left(-\frac{\zeta^2}{4D}\right),$$ (5.32)

where C is an arbitrary constant.

From Eq. (5.32) it follows that

$$\hat{c}(\xi, \tau) = C \exp\left(-\frac{\xi^2}{4D\tau}\right).$$

We determine the constant C by imposing the condition (5.26), getting

$$\hat{c}(\xi, \tau) = \frac{1}{\sqrt{4\pi D\tau}} \exp\left(-\frac{\xi^2}{4D\tau}\right) = \hat{c}_\delta(\xi, \tau).$$ (5.33)

Therefore, the solution to the moving plume problem is

$$c(x, t) = \frac{1}{\sqrt{4\pi Dt}} \exp\left[\frac{-(x - vt)^2}{4Dt}\right].$$

Figure 5.6 shows graphs of $c(x, t)$, at two different values of time t.

The solution $c_\delta(\xi, \tau)$ given by (5.33) is the **fundamental solution** to the heat equation. There are other ways to derive this solution; see, for example, [65, Section 5-4]. From a heuristic viewpoint, the fundamental solution gives the response to an initial, unit-strength point source centered at spatial position 0. Because the heat equation (5.24) is linear, this heuristic suggests a **superposition principle** for obtaining responses to more complicated initial conditions: Superpose weighted responses to unit-strength point sources centered at points where the initial condition is nonzero. Section 5.3.2 exploits this principle.

5.3.2 The Moving Front Problem

We now examine a second initial-value problem for Eq. (5.24), the **moving front** problem. Here the initial condition is

$$\hat{c}(\xi, 0) = c_I(\xi) = \begin{cases} 1, & \text{if } \xi \leq 0, \\ 0, & \text{if } \xi > 0. \end{cases}$$ (5.34)

Figure 5.7 Graph of the fundamental solution to the heat equation showing the area represented by the integral in Eq. (5.35).

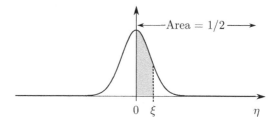

This condition models a constant source of solute at $\xi = 0$, which we regard heuristically as a continuum of unit-strength point sources in the interval $\xi \leq 0$.

Superposition of the responses to these point sources yields

$$\hat{c}(\xi, \tau) = \underbrace{\int_{-\infty}^{\infty} c_I(y)}_{(I)} \ \underbrace{\hat{c}_\delta(\xi - y, \tau)}_{(II)} \, dy. \tag{5.35}$$

In this integral, the factor (I), obtained from the initial condition (5.34), represents the weighting factor as a function of position y in the domain of integration. The fundamental solution (II), defined in Eq. (5.33), gives the response at time τ to a unit-strength point source $\delta(\xi - y)$ centered at spatial position $y = \xi$.

Using Figure 5.7 to calculate the integral in Eq. (5.35) yields

$$\hat{c}(\xi, \tau) = \int_{-\infty}^{0} \hat{c}_\delta(\underbrace{\xi - y}_{\eta}, \tau) \, dy = \int_{\xi}^{\infty} \hat{c}_\delta(\eta, \tau) \, d\eta = \frac{1}{2} - \int_{0}^{\xi} \hat{c}_\delta(\eta, \tau) \, d\eta$$

$$= \frac{1}{2} - \int_{0}^{\xi} \frac{1}{\sqrt{4\pi D\tau}} \exp\left(-\frac{\eta^2}{4D\tau}\right) d\eta$$

$$= \frac{1}{2} - \int_{0}^{\xi/\sqrt{4D\tau}} \frac{1}{\sqrt{\pi}} e^{-y^2} \, dy = \frac{1}{2} - \frac{1}{2}\mathrm{erf}\left(\frac{\xi}{\sqrt{4D\tau}}\right).$$

Here,

$$\mathrm{erf}\,(u) = \frac{2}{\sqrt{\pi}} \int_{0}^{u} e^{-y^2} \, dy$$

is the **error function**. Substituting for ξ and τ gives

$$c(x, t) = \frac{1}{2} - \frac{1}{2}\mathrm{erf}\left(\frac{x - vt}{\sqrt{4Dt}}\right). \tag{5.36}$$

Figure 5.8 shows graphs of this solution, at a fixed time $t > 0$, for several values of the diffusion coefficient D. Two facts are apparent:

1. For positive values of D, the sharp front at $x = 0$ in the initial condition spreads out to a smooth solution as t increases, with positive values of $c(x, t)$ instantaneously appearing at all values of x. (As $x \to \infty$, these values decay to 0 rapidly.)

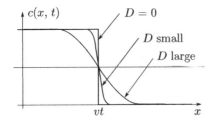

Figure 5.8 Graph of moving front solutions (5.36) to the advection–diffusion equation for different values of the diffusion coefficient D.

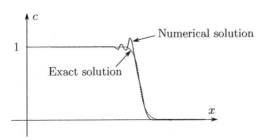

Figure 5.9 Snapshot of a numerical solution to the moving-front problem using a centered-in-space finite-difference approximation to the advection–diffusion equation.

This observation confirms the infinite propagation speed that we associate with the heat equation in Section 4.1.

2. As D decreases, the moving front, centered at $x = vt$, remains steep for longer times. The case $D \to 0$ corresponds to the pure advection equation (5.15).

Fluid-flow problems whose solutions have steep fronts confront numerical modelers with vexing challenges. At the root of the problem is the use of discrete grids to approximate spatial derivatives: It is impossible to resolve a front whose width is much smaller than the width h of a cell in the computational grid.

This fundamental difficulty manifests itself in several ways, depending on the type of numerical approximations employed. For example, finite-difference or finite-element discretizations that possess high-order spatial accuracy, as measured by the truncation errors of the derivative approximations, tend to produce solutions that exhibit spurious oscillations near steep fronts. Figure 5.9 shows a snapshot in time of a typical numerical solution of the one-dimensional advection–diffusion equation using a method in which the spatial truncation error is $\mathcal{O}(h^2)$, where h is the grid spacing. This approximate solution captures the sharp front, but it oscillates near the front, in contrast to the true solution. American engineer William G. Gray and Canadian-American geohydrologist George F. Pinder [63] have shown that the spurious numerical oscillations arise from numerically induced errors in the propagation speeds of the short-wavelength Fourier modes needed to resolve the sharp front.

On the other hand, finite-difference or finite-element discretizations that possess low-order spatial accuracy, such as upstream weighted approximations, introduce truncation errors that add artificial numerical diffusion. This effect

Figure 5.10 Snapshot of a numerical solution to the moving-front problem using an upstream-weighted finite-difference approximation to the advection–diffusion equation.

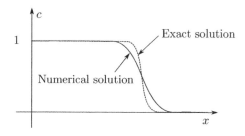

eliminates spurious oscillations at the expense of excessive smearing of the sharp front. Figure 5.10 shows a snapshot in time of a typical approximate solution of the one-dimensional advection–diffusion equation, using a backward-in-space finite-difference method in which the spatial truncation error is $\mathcal{O}(h)$. For this numerical approximation, the lowest-order term in the truncation error has a diffusion-like form; see [5, Section 9.4.2] for an explanation.

Numerical modelers of advection-dominated transport in porous media must navigate between the Scylla of spurious oscillations and the Charybdis of numerical diffusion. In some applications, the need to capture sharp fronts outweighs the problems associated with oscillations, and high-order spatial approximations are appropriate. However, if the numerically induced oscillations trigger additional effects, such as threshold-limited chemical reactions, then low-order spatial approximations may yield more realistic model results, despite the artificial smearing. In some nonlinear problems, such as those explored in Section 6.4, some form of numerical diffusion is necessary to obtain physically realistic results. One way to manage either type of error is to use adaptive local grid refinement algorithms, which can control the numerical errors near sharp moving fronts by reducing the sizes of the spatial grid cells in a small region around the fronts. For details, see [5, Chapter 9].

5.4 Transport with Adsorption

Many flows in porous media involve exchanges of mass between phases. Consider transport of a solute, having species index i, in a fluid phase F flowing through a porous medium. If the solute adsorbs onto the rock phase R, then we must account for the mass of species i—which we call the **adsorbate**, regardless of the phase in which it resides—using mass balance equations for constituents (i, F) and (i, R). To simplify notation, throughout this section we denote the concentrations of these constituents as follows:

$$c_{(i,F)}(\mathbf{x}, t) = c(\mathbf{x}, t) \qquad \text{Concentration in fluid,}$$
$$c_{(i,R)}(\mathbf{x}, t) = a(\mathbf{x}, t) \qquad \text{Concentration in rock.}$$

5.4.1 Mass Balance for Adsorbate

If the rock is stationary and the adsorbate does not diffuse in the rock, then the mass balance equations for species i in the phases F and R are as follows:

$$\frac{\partial}{\partial t}(\phi c) = -\nabla \cdot (\phi \mathbf{v}c) + \nabla \cdot (\phi D \nabla c) + r_{(i,F)},$$

$$\frac{\partial}{\partial t}[(1 - \phi)a] = r_{(i,R)}, \tag{5.37}$$

respectively. If no chemical reactions produce or consume the adsorbate, then $r_{(i,F)} + r_{(i,R)} = 0$, so adding Eqs. (5.37) gives the total adsorbate mass balance,

$$\frac{\partial}{\partial t}[\phi c + (1 - \phi)a] = -\nabla \cdot (\phi \mathbf{v}c) + \nabla \cdot (\phi D \nabla c). \tag{5.38}$$

In many applications, adsorption is a fast reaction. In these cases, we typically neglect the reaction kinetics and assume that the constituents (i, F) and (i, R) reach a concentration-dependent equilibrium instantaneously, so that $a = a(c)$. This equilibrium relationship is called an **adsorption isotherm**, since it is typically valid for a fixed temperature. Incorporating this model into the adsorbate mass balance (5.38) yields

$$\frac{\partial}{\partial t}[\phi c + (1 - \phi)a(c)] = -\nabla \cdot (\phi \mathbf{v}c) + \nabla \cdot (\phi D \nabla c). \tag{5.39}$$

Among the commonly used models of adsorption isotherms are the following:

- The **linear isotherm**, which has the form $a(c) = \kappa c$ for a positive, empirically determined constant κ.
- The **Freundlich isotherm**, named for German chemist Herbert Freundlich and having the form $a(c) = \kappa c^n$, where κ and n are positive, empirically determined constants. The utility of this model arises not from any theoretical foundation but from its amenability to experimental curve fitting.
- The **Langmuir isotherm**, named for American chemist Irving Langmuir and having the form

$$a(c) = \frac{\kappa_1 c}{1 + \kappa_2 c}, \tag{5.40}$$

for positive constants κ_1, κ_2. This form enjoys some theoretical foundation; see [92]. Nevertheless, in practice one must determine the parameters κ_1, κ_2 empirically. In contrast to the linear and Freundlich isotherms, the Langmuir isotherm has a horizontal asymptote that represents the saturated state of the rock:

$$\lim_{c \to \infty} a(c) = \frac{\kappa_1}{\kappa_2}.$$

Figure 5.11 (a) Linear isotherm. (b) Freundlich isotherms for different values of the positive exponent n. (c) Langmuir isotherm showing the asymptotic value κ_1/κ_2.

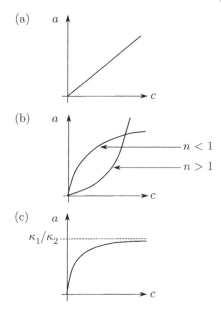

As Figure 5.11 shows, the graph of the Langmuir isotherm is concave down, that is, $a''(c) < 0$, while the concavity of the Freundlich isotherm depends on the choice of exponent n. One can regard the linear isotherm as a reasonable approximation to the Langmuir isotherm when the adsorbate concentration c in the fluid is small.

5.4.2 Linear Isotherms and Retardation

Simple analyses lend insight into the effects that the linear and Langmuir isotherms exert on adsorbate transport. Consider first the linear isotherm. If the porosity is constant and the fluid flow is incompressible, then $\nabla \cdot (\phi \mathbf{v}) = 0$, and the total adsorbate mass balance (5.38) becomes

$$\frac{\partial}{\partial t}[\phi + \kappa(1 - \phi)]c = -\phi \mathbf{v} \cdot \nabla c + \nabla \cdot (\phi D \nabla c). \tag{5.41}$$

When ϕ and κ are constants, we can simplify this equation even further by defining a **retardation factor**

$$R = 1 + \frac{\kappa(1 - \phi)}{\phi}.$$

Since $\kappa > 0, R > 1$. Substituting this factor into Eq. (5.41) and dividing by ϕR yields

$$\frac{\partial c}{\partial t} + \frac{\mathbf{v}}{R} \cdot \nabla c - \nabla \cdot \left(\frac{D}{R} \nabla c \right) = 0. \tag{5.42}$$

Equation (5.42) has the same form as the advection–diffusion equation (5.12) in the absence of adsorption, with the advection and diffusion coefficients replaced by retarded transport coefficients $\mathbf{v}/R$ and D/R. Thus, the effect of adsorption in the case of a linear adsorption isotherm is to cause adsorbate plumes and fronts to advect and diffuse more slowly than they would in the absence of adsorption.

5.4.3 Concave-down Isotherms and Front Sharpening

The Langmuir isotherm has more exotic effects. These are easiest to analyze in the one-dimensional version of the total adsorbate mass balance (5.39) in which the fluid velocity is constant and hydrodynamic dispersion is negligible:

$$\frac{\partial}{\partial t}[\phi c + (1 - \phi)a(c)] + v\frac{\partial c}{\partial x} = 0. \tag{5.43}$$

If the porosity ϕ is constant, then the chain rule yields

$$[\phi + (1 - \phi)a'(c)]\frac{\partial c}{\partial t} + v\frac{\partial c}{\partial x} = 0. \tag{5.44}$$

We now apply the method of characteristics discussed in Section 5.2. If the concentration $c(x, t)$ is sufficiently smooth, then along any continuously differentiable path $(x(s), t(s))$ in the (x, t)-plane with nonzero derivative $(x'(s), t'(s))$, the chain rule requires

$$t'(s)\frac{\partial c}{\partial t} + x'(s)\frac{\partial c}{\partial x} = \frac{dc}{ds}. \tag{5.45}$$

We seek paths along which this directional derivative is identical to the left side of Eq. (5.44), that is, paths for which

$$t'(s) = \phi + (1 - \phi)a'(c), \quad x'(s) = v.$$

Equivalently,

$$\frac{x'(s)}{t'(s)} = \frac{dx}{dt} = U(c), \tag{5.46}$$

where

$$U(c) = \frac{v}{\phi + (1 - \phi)a'(c)}. \tag{5.47}$$

Along these characteristic curves, the right sides of Eqs. (5.43) and (5.45) must agree. Thus, the concentration c obeys the characteristic equation

$$\frac{dc}{ds} = 0 \tag{5.48}$$

along characteristic curves.

This result has two consequences. The first arises from the observation that loci of constant concentration travel with speed $U(c)$ given by Eqs. (5.46) and (5.47). Since $0 < \phi < 1$, $U(c) < v$ whenever $a'(c) > 1$. Under this condition, Langmuir adsorption retards the advective transport of the adsorbate.

Figure 5.12 Graph of the ramp-like initial concentration profile for the initial-value problem (5.49).

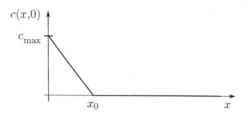

The second, more interesting consequence arises from the fact that $a''(c) < 0$. This inequality implies that solute concentration fronts tend to sharpen in time. To understand this effect, assume that we have a concentration profile prescribed along an initial curve in the (x, t)-plane. Each of the characteristic curves defined by Eq. (5.46) is a straight line, whose slope $U(c)$ depends on the value of c where the line intersects the initial curve.

Exercise 5.13 *Use the fact that $a''(c) < 0$ to show that $U(c)$ in Eq. (5.47) is an increasing function of c.*

Consider, for example, the ramp-like initial-boundary condition

$$c(0, t) = c_{max}, \quad t \geq 0;$$

$$c(x, 0) = \begin{cases} c_{max} - (c_{max}/x_0)x, & \text{if } 0 < x \leq x_0, \\ 0, & \text{if } x > x_0, \end{cases} \tag{5.49}$$

drawn in Figure 5.12. For this Cauchy problem, the initial curve is the union of the positive x-axis with the nonnegative t-axis.

Since the speed $U(c)$ increases with c by Exercise 5.13, the characteristic curves for the initial-boundary conditions (5.49), having slopes $dx/dt = U(c)$, must intersect at some finite time t_s, as shown in Figure 5.13. But Eq. (5.48) requires the value of the concentration c to remain constant along characteristic curves, so allowing the characteristic curves to cross would produce a multivalued—and hence physically unrealistic—solution, starting at time t_s.

The resolution, suggested in Figure 5.14, is to allow the solution $c(x, t)$ to develop a jump discontinuity, or **shock**. Such a solution is necessarily a weak solution, as discussed in Section 5.2, since the PDE (5.43) cannot hold in the strict sense at a discontinuity. One can guarantee a well-defined solution by choosing one of the characteristic curves emanating from the interval $[0, x_0]$ on the x-axis in Figure 5.13 to serve as the boundary, for $t \geq t_s$, between the region where $dx/dt = U(c_{max})$ and the region where $dx/dt = U(0)$. But there are infinitely many such characteristic curves. Simply requiring the solution to be single-valued does not suffice to define $c(x, t)$ uniquely.

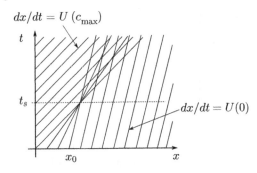

$dx/dt = U\,(c_{max})$

$dx/dt = U(0)$

Figure 5.13 Characteristic curves associated with the initial-value problem (5.49) showing that a concave-down adsorption isotherm causes characteristic curves to intersect.

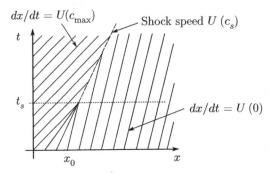

$dx/dt = U(c_{max})$

Shock speed $U\,(c_s)$

$dx/dt = U\,(0)$

Figure 5.14 Resolution to the problem of intersecting characteristics in Figure 5.13 by allowing a shock moving with speed $U(c_s)$ determined using an integral form of the total adsorbate mass balance.

5.4.4 The Rankine–Hugoniot Condition

To choose the correct locus of the shock in the (x, t)-plane, we must find a shock speed dx/dt that respects the physics of the problem. For this purpose, assume that the solution $c(x, t)$ is continuously differentiable everywhere except at a jump discontinuity located at a point $x = \Sigma(t)$, to be determined, for $t \geq t_s$. The PDE

$$\frac{\partial A}{\partial t}(c) + v\frac{\partial c}{\partial x} = 0, \quad \text{where } A(c) = \phi c + (1 - \phi)a(c),$$

no longer holds throughout any subinterval $[x_L, x_R]$ of the spatial domain that contains $\Sigma(t)$, since the function $c(x, t)$ is no longer differentiable everywhere in that interval. To calculate the shock speed, we develop a version of the mass balance equation that relaxes the smoothness requirements on $c(x, t)$.

To make the calculation useful in later problems as well as the current one, let us examine a more general **conservation law** having the form

$$\frac{\partial A}{\partial t}(c) + \frac{\partial F}{\partial x}(c) = 0, \tag{5.50}$$

where $A(c)$ denotes an accumulation function and $F(c)$ represents a generic **flux function**. For the current problem,

$$A(c) = \phi c + (1 - \phi)a(c); \quad F(c) = vc. \tag{5.51}$$

Exercise 5.14 *Show that, for any fixed interval $[x_L, x_R]$ in which Eq. (5.50) holds,*

$$\frac{d}{dt} \int_{x_L}^{x_R} A \, dx + F(c(x_R, t)) - F(c(x_L, t)) = 0. \tag{5.52}$$

Equation (5.52) says that the net flux of c at the endpoints x_L and x_R exactly balances the rate of accumulation in the interval. This **integral conservation law** holds under more general conditions than the differential equation (5.50), since the former does not assume that $c(x, t)$ is differentiable in the interval $[x_L, x_R]$.

Now examine the integral in Eq. (5.52) more closely, by splitting it at the yet-to-be-determined locus $\Sigma(t)$ of the jump discontinuity:

$$\frac{d}{dt} \int_{x_L}^{\Sigma(t)} A \, dx + \frac{d}{dt} \int_{\Sigma(t)}^{x_R} A \, dx + F(c(x_R, t)) - F(c(x_L, t)) = 0.$$

By the Leibniz rule,

$$\frac{d}{dt} \int_{x_L}^{\Sigma(t)} A \, dx = \int_{x_L}^{\Sigma(t)} \frac{\partial A}{\partial t} \, dx + A(c(\Sigma-, t)) \, \Sigma'(t),$$

$$\frac{d}{dt} \int_{\Sigma(t)}^{x_R} A \, dx = \int_{\Sigma(t)}^{x_R} \frac{\partial A}{\partial t} \, dx - A(c(\Sigma+, t)) \, \Sigma'(t),$$

where

$$c(\Sigma-, t) = \lim_{x \to \Sigma(t)-} c(x, t), \quad c(\Sigma+, t) = \lim_{x \to \Sigma(t)+} c(x, t).$$

Upon combining the surviving integrals, we find that

$$\int_{x_L}^{x_R} \frac{\partial A}{\partial t} \, dx - [\![b]\!] \Sigma'(t) + F(c(x_R, t)) - F(c(x_L, t)) = 0, \tag{5.53}$$

where

$$[\![\cdot]\!] = \lim_{x \to \Sigma+} (\cdot) - \lim_{x \to \Sigma-} (\cdot)$$

denotes the **jump** in any variable $(\cdot)$ across the shock.

Now let $x_L \to \Sigma-$ and $x_R \to \Sigma+$. In this limit the integral in Eq. (5.53) vanishes, leaving the identity

$$\boxed{\Sigma'(t) = \frac{[\![F]\!]}{[\![A]\!]}.} \tag{5.54}$$

Equation (5.54) is the **Rankine–Hugoniot condition** for the shock speed $\Sigma'(t)$. It is a variant of the original conservation law (5.50) valid at a jump discontinuity in the solution $c(x, t)$.

For the accumulation and flux functions (5.51) of interest in the adsorption problem, the Rankine–Hugoniot condition yields the following equation:

$$\Sigma'(t) = \frac{v [\![c]\!]}{[\![A]\!]} = \frac{v}{\phi + (1 - \phi) [\![a]\!] / [\![c]\!]}. \tag{5.55}$$

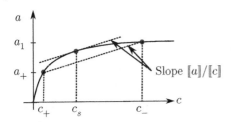

Figure 5.15 Chord on the isotherm defining the speed of the adsorbate concentration shock. Also shown is the concentration value c_s guaranteed by the mean value theorem.

As Figure 5.15 illustrates, the ratio $[\![a]\!]/[\![c]\!]$ that appears in the denominator represents the slope of the chord connecting the points on the isotherm corresponding to upstream and downstream concentration values, c_- and c_+, respectively, at the shock $\Sigma(t)$. By the mean value theorem, there exists a concentration value c_s between c_- and c_+ such that

$$a'(c_s) = \frac{[\![a]\!]}{[\![c]\!]}. \tag{5.56}$$

Therefore, utilizing the definition (5.47), we can write Eq. (5.55) for the shock speed as

$$\Sigma'(t) = U(c_s).$$

This equation gives the slope of the characteristic curve that separates the region in Figure 5.14 where $dx/dt = U(c_-)$ from the region where $dx/dt = U(c_+)$.

For the initial-value problem (5.49), $U(c_-) = U(c_{max})$ and $U(c_+) = U(0)$. The solution $c(x, t)$, plotted at several values of t, appears in Figure 5.16. This solution remains continuous until the larger concentration values overtake the smaller ones at time t_s. After that, a shock separating the value c_{max} from 0 propagates to the right with a speed $U(c_s)$ whose value lies between those associated with the two endpoint concentrations.

Exercise 5.15 *Using the normalized concentration value $c_{max} = 1$, the Langmuir isotherm (5.40), and Eq. (5.56), calculate c_s for the initial-boundary conditions (5.49) in terms of the parameters κ_1 and κ_2. Use the result to calculate the shock speed $U(c_s)$.*

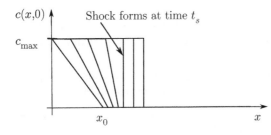

Figure 5.16 Graphs of the solution $c(x, t)$ at several time levels, showing the formation of a concentration shock under the influence of a concave-down adsorption isotherm.

In reality, diffusion smooths the sharp concentration profile. Nevertheless, the method of characteristics, applied in the diffusion-free limit, reveals that any concave-down adsorption isotherm exerts a sharpening—and hence numerically vexing—effect on advancing concentration fronts. A similar front-sharpening effect occurs in multifluid flows in porous media, as Chapter 6 explores.

6

Multifluid Flows

Complexities arise when several immiscible fluid phases flow through the rock. The feature that distinguishes these flows from single-fluid flows is the presence of at least one continuum-scale interface between different fluid phases. Although we can detect this interface through microscopes, we typically do not see it or explicitly model it at scales of observation larger than around 10^{-3} m. This fact notwithstanding, the physics of microscopic fluid–fluid interfaces exert profound influences on the macroscopically observable properties of these flows, so Section 6.1 devotes some attention to the microscopic scale.

Flows of this type occur in a wide range of settings, including:

- Petroleum reservoirs (oil + gas + brine)
- Contaminated aquifers (water + nonaqueous-phase liquids, NAPL)
- Near-surface soils (water + air, possibly with NAPL)
- Carbon dioxide sequestration (brine + CO_2)
- Geothermal reservoirs (liquid water + steam).

Far more exotic fluid mixtures appear in applications such as enhanced oil recovery and groundwater remediation, where various surfactants, foams, and emulsions flow simultaneously with other fluid phases. In these flows, chemical reactions and interphase mass transfer play important roles in the design. Although the balance laws established in Chapter 2 can accommodate these more complicated physics, this chapter restricts attention to multifluid flows in which chemical reactions and interphase mass transfer are absent or negligible. Chapter 7 introduces models of flows with interphase mass transfer.

Throughout this chapter, for simplicity, we treat the permeability as isotropic. One can relax this assumption; see Section 3.7.

The Mathematics of Fluid Flow Through Porous Media, First Edition. Myron B. Allen.
© 2021 John Wiley & Sons, Inc. Published 2021 by John Wiley & Sons, Inc.

6.1 Capillarity

6.1.1 Physics of Curved Interfaces

Every interface between immiscible fluids possesses surface energy. We define the **interfacial tension** σ between two immiscible fluids to be the energy per unit area required to maintain the two-dimensional interface that separates them. For any pair of chemically homogeneous, immiscible fluids at a fixed temperature, we treat σ as a positive constant. (When the fluids are miscible, $\sigma = 0$.) The term interfacial *tension* reflects the fact that the dimension MT^{-2} of energy per unit area is the same as the dimension of force per unit length. In this interpretation, σ gives the magnitude of the force per unit area tangent to the interface, acting perpendicular to any arc lying in it. This section exploits this interpretation in analyzing the balance of forces on an interface.

The following principle lies at the core of multifluid flow through porous media: If the interface between two immiscible fluids is curved, then the fluids have different pressures at the interface. A mathematical explanation of this principle, reviewed in this subsection, requires the use of integrals along paths and integrals over surfaces. A brief review of integrals over smooth surfaces appears in Appendix D.

To see how interfacial tension affects fluid pressures, consider any smooth piece Σ of the surface separating immiscible fluids labeled 1 and 2. Denote the curve bounding the surface by $\partial\Sigma$, as drawn in Figure 6.1. Assume the following:

- Σ has a continuously differentiable parametrization (see Appendix D) and therefore possesses a unit-length normal vector field $\mathbf{n}(\mathbf{x})$ that points toward fluid 2.
- There exists a continuously differentiable, closed path γ, defined on a parameter interval $[a, b]$, that parametrizes the boundary $\partial\Sigma$, with $\|\gamma'(s)\| = 1$ for all $s \in [a, b]$.
- The path γ is oriented positively with respect to the normal vector field $\mathbf{n}$. In other words, as we look down on the surface from the direction toward which $\mathbf{n}$ points, γ traces $\partial\Sigma$ counterclockwise, as shown in Figure 6.1.

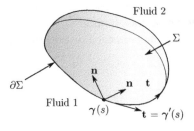

Figure 6.1 A piece Σ of smooth surface with boundary $\partial\Sigma$, which is parametrized by a smooth path γ. At each point $\gamma(s)$ on the path, the vector $\mathbf{t}(s) = \gamma'(s)$ is tangent to the path and to the surface; $\mathbf{n}(s)$ is perpendicular to the surface, and $\mathbf{n}(s) \times \mathbf{t}(s)$ is also tangent to the surface but perpendicular to the path.

For each parameter value $s \in [a, b]$, the unit-length vector $\mathbf{t}(s) = \gamma'(s)$ is tangent to the path—and hence to the surface Σ—at the point $\gamma(s)$ on $\partial\Sigma$. Also tangent to Σ at $\gamma(s)$ is the unit-length vector $\mathbf{n}(\gamma(s)) \times \mathbf{t}(s)$, which is perpendicular to the path at $\gamma(s)$. In this geometry, the force perpendicular to γ, per unit length of arc, at the point $\gamma(s)$ is $\sigma \, \mathbf{n}(\gamma(s)) \times \mathbf{t}(s)$. The total force exerted by interfacial tension along the arc γ is therefore

$$\int_\gamma \sigma \, \mathbf{n} \times \mathbf{t} \, ds = \int_a^b \sigma \, \mathbf{n}(\gamma(s)) \times \gamma'(s) \, ds.$$

Denote the pressure of fluid phase 1 by p_1 and that of fluid phase 2 by p_2. These pressures give the outward force per unit area acting perpendicular to the surface Σ. If the interface is at equilibrium, these forces on Σ balance the interfacial force:

$$\int_\Sigma p_1 \mathbf{n} \, da + \int_\Sigma p_2(-\mathbf{n}) \, da - \int_\gamma \sigma \mathbf{n} \times \mathbf{t} \, ds = \mathbf{0}.$$

Here da denotes the element of surface integration, reviewed in Appendix D.

To simplify this equation, we convert the boundary integral over the path γ to a surface integral over Σ. This task uses Theorem D.3.1, a corollary to the Stokes theorem, also described in Appendix D, with $\sigma\mathbf{n}$ playing the role of the function $\mathbf{f}$ in that theorem:

$$\int_\gamma \sigma\mathbf{n} \times \mathbf{t} \, ds = \int_\Sigma \sigma \, [(\nabla \cdot \mathbf{n})\mathbf{n} - \mathbf{n} \cdot (\nabla\mathbf{n})] \, da.$$

See Eq. (D.2) for the representation of $\mathbf{n} \cdot \nabla\mathbf{n}$ in Cartesian coordinates.

Exercise 6.1 *Use the product rule and the fact that $\mathbf{n}$ is a unit-length vector field to show that $\mathbf{n} \cdot (\nabla\mathbf{n}) = \mathbf{0}$.*

It follows from Exercise 6.1 that only the first term in the integral on the right survives:

$$\int_\Sigma (p_1 - p_2 - \sigma\nabla \cdot \mathbf{n})\mathbf{n} \, da = \mathbf{0}. \tag{6.1}$$

Exercise 6.2 *Justify the assertion that the only way Eq. (6.1) can hold for arbitrary pieces Σ of smooth surface is for the integrand to vanish:*

$$p_1 - p_2 = \sigma\nabla \cdot \mathbf{n}. \tag{6.2}$$

Equation (6.2), a local force balance, is the **Young–Laplace equation**, named for the English physician Thomas Young and the renowned French mathematician Pierre-Simon Laplace.

Exercise 6.3 *Calling Thomas Young a physician scarcely does him justice. Review a biography to learn how many fields this early nineteenth-century polymath advanced through his original contributions. For a detailed biography, see* [130].

As a force balance, Eq. (6.2) admits further simplification. Differential geometers identify the function $-\frac{1}{2}\nabla \cdot \mathbf{n}(\mathbf{x})$ as the **mean curvature** of Σ at $\mathbf{x}$ [147, p. 94]. It measures the rate of change, with respect to position, of the unit normal vector $\mathbf{n}(\mathbf{x})$ at each point $\mathbf{x} \in \Sigma$. To calculate this function, we represent Σ as a level set $F(\mathbf{x}) = 0$ of some continuously differentiable, real-valued function F for which $\nabla F(\mathbf{x})$ never vanishes. Since the vector $\nabla F(\mathbf{x})$ is orthogonal to the level set at $\mathbf{x}$, we can define a unit-length normal vector field for Σ as follows:

$$\mathbf{n}(\mathbf{x}) = \frac{\nabla F(\mathbf{x})}{\|\nabla F(\mathbf{x})\|}, \quad \text{for } \mathbf{x} \in \Sigma.$$

From this expression, one can calculate $-\frac{1}{2}\nabla \cdot \mathbf{n}(\mathbf{x})$ explicitly.

In this calculation, the direction of $\mathbf{n}(\mathbf{x})$ depends on the choice of the function F; hence, so does the sign of $-\frac{1}{2}\nabla \cdot \mathbf{n}(\mathbf{x})$. At points $\mathbf{x} \in \Sigma$ where the interface bends toward $\mathbf{n}(\mathbf{x})$ in all directions—that is, points where Σ is concave when viewed from the side toward which $\mathbf{n}$ points—the mean curvature is positive. At points where the interface bends away from $\mathbf{n}(\mathbf{x})$—where Σ is convex when viewed from the side toward which $\mathbf{n}$ points, as in Figure 6.1—the mean curvature is negative.

Exercise 6.4 *Verify the mean curvatures of the two constant-curvature surfaces listed below by representing each as a level set of some function $F(\mathbf{x})$.*

1. *A sphere of radius R, with $\mathbf{n}$ pointing away from the center of the sphere, has mean curvature $-1/R$.*
2. *The side of a circular cylinder of radius R, with $\mathbf{n}$ pointing away from the axis has mean curvature $-1/(2R)$.*

Exercise 6.5 *Sketch the surface defined by the equation $F(x_1, x_2, x_3) = x_1 x_2 - x_3 = 0$, then calculate the mean curvature at $\mathbf{x} = (0, 0, 0)$.*

By analogy with the result for the mean curvature of the sphere, we define the **mean radius of curvature** $\bar{r}$ of a surface having unit normal vector field $\mathbf{n}(\mathbf{x})$ as follows:

$$\bar{r} = \left| \frac{1}{\frac{1}{2}\nabla \cdot \mathbf{n}} \right|.$$

Thus, if the mean curvature is negative, as illustrated in Figure 6.1, then $1/\bar{r} = \frac{1}{2}\nabla \cdot \mathbf{n}$. With this notation, the Young–Laplace equation (6.2) becomes

$$\boxed{p_C = p_1 - p_2 = \frac{2\sigma}{\bar{r}},}$$

provided p_1 is the pressure of the fluid on the concave side of the interface.

Figure 6.2 A curved interface Σ with the pressure on the concave side being greater than the pressure on the convex side.

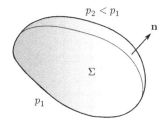

$$p_2 < p_1$$

We call the pressure difference $p_C = p_1 - p_2$ the **capillary pressure** between the fluids 1 and 2. In particular, for any curved interface between two immiscible fluids (Figure 6.2):

> The pressure on the concave side is greater than the pressure on the convex side.

6.1.2 Wettability

The Young–Laplace equation (6.2) bears on fluid mechanics in porous media because all immiscible fluid–fluid interfaces in the pore space are curved. In the presence of two immiscible fluids, any solid surface exhibits a preferential affinity for one fluid over the other [103]. This affinity is manifested by a **contact angle** θ_c between the two fluids where their interface meets the solid surface, as sketched in Figure 6.3 for two fluids in a solid tube. We call the fluid in which θ_c is an acute angle the **wetting fluid** W; the fluid on the other side of the interface is the **nonwetting fluid** N. By convention, $p_C = p_N - p_W$.

To grasp the influence that wettability and capillary pressure exert on fluids in a porous medium, consider a single cylindrical glass tube having radius R, with one end inserted into water and the other open to the air above it, as drawn in Figure 6.4. In this simple physical model of a channel in a porous medium, water is typically the wetting fluid W, with an acute contact angle θ_c. The resulting curvature of the air–water interface results in a pressure difference according to the Young–Laplace equation. As a consequence, the equilibrium height of water in

Figure 6.3 Wetting fluid W and nonwetting fluid N in a tube with contact angle θ_c between the wetting fluid and the solid wall.

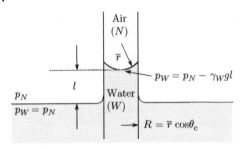

Figure 6.4 Capillary rise of water in a tube open at the top to the air.

the tube differs from the height of water outside the tube. To relate the height difference l to the radius R of the tube, observe that, at equilibrium, the hydrostatic equation (2.13) implies that

$$p_N = p_W + \gamma_W g l.$$

Here, the index N signifies the air phase, and γ_W denotes the true density of water. Thus, by definition, $p_C = p_N - p_W = \gamma_W g l$. But by the Young–Laplace equation, $p_C = 2\sigma/\bar{r} = (2\sigma \cos\theta_c)/R$. Equating the two expressions for p_C gives **Jurin's law**:

$$l = \frac{2\sigma \cos\theta_c}{\gamma_W g}\frac{1}{R},$$

named after the English scientist and physician James Jurin, who examined capillary rise in tubes in 1718 [85]—one of the earliest studies in porous-flow physics. In short, the **capillary rise** is inversely proportional to the radius of the tube.

For further insight into the effects of capillarity on multifluid flows in porous media, consider a slightly more complex model, namely a bundle of glass tubes having random radii, as shown in Figure 6.5. Here we see a range of values of capillary rise, reflecting the statistics of the radii. This idealized analog of a porous medium suggests that a similar range of microscopic fluid levels will characterize the rise of the wetting fluid in the tortuous and irregular pore space of a natural porous medium. A simple way to characterize the distribution of water levels is to

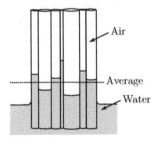

Figure 6.5 Capillary rise of water in a bundle of tubes having different radii.

specify, for the total volume of the tubes standing above the external water level, the fraction of the volume occupied by water. This fraction, which the Section 6.1.3 generalizes as water saturation, provides a gauge of the average capillary pressure in the tubes.

6.1.3 Capillarity at the Macroscale

Capillary bundles provide a conceptual bridge to the macroscopic scale, appropriate to the mixture model of porous media. American engineer Miles C. Leverett [96] pursued this line of reasoning in 1941; see [52, 74], and [138] for examples of more recent work.

Consider a three-phase continuum consisting of immiscible phases R (rock), W (wetting fluid), and N (nonwetting fluid). For each fluid $\alpha = N, W$, define the **saturation** as $S_\alpha = \phi_\alpha/\phi$, where ϕ_α denotes the volume fraction of phase α and, as in Chapter 3, $\phi = 1 - \phi_R$ is the porosity of the rock. Assume that the pore space is completely filled by N and W, so $\phi = \phi_N + \phi_W$. It follows that

$$S_N + S_W = 1.$$

At this scale, we regard the macroscopic capillary pressure p_C as an average of capillary pressure values taken over a representative elementary volume (see Section 2.5) in the porous medium. The reasoning at the end of Section 6.1.2 suggests the simple hypothesis $p_C = p_C(S_W)$, a function that one must measure experimentally to characterize the particular system of rock and fluids being modeled.

In reality, this commonly adopted hypothesis is too simple. When plotted against S_W, carefully measured values of p_C typically depend not only on the rock and fluids being tested but also on the sample's saturation history. In particular, the values of p_C depend on whether S_W is increasing or decreasing. In 1930, British soil scientist William B. Haines [68] examined this effect and proposed microgeometric mechanisms to explain these history dependencies. For present purposes it suffices to observe that experimental measurements of p_C versus S_W give rise to curves that exhibit **hysteresis**, that is, the values of p_C follow different curves depending on the saturation history.

Figure 6.6 illustrates this effect, along with several other important features of a typical capillary-pressure plot. The term **imbibition** refers to flows in which the saturation S_W of the wetting fluid increases, while the term **drainage** refers to flows in which S_W decreases. Because of the rock's greater affinity to the wetting fluid, drainage requires more energy—and hence higher values of capillary pressure at a given value of S_W—than imbibition.

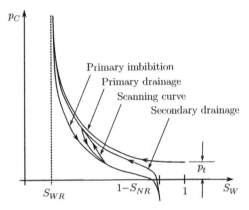

Figure 6.6 Typical capillary pressure curves showing the effects of hysteresis. The drainage curves characterize flows in which the wetting-fluid saturation decreases. The imbibition curves characterize flows in which the wetting-fluid saturation increases.

In Figure 6.6, the **primary drainage curve** starts with a medium completely saturated with the wetting fluid. As the plot indicates, displacement of the resident wetting fluid by nonwetting fluid at $S_W = 1$ requires a minimum value of p_C, called the **threshold pressure**, denoted by p_t in Figure 6.6. As S_W decreases, p_C increases toward a vertical asymptote at a saturation value $S_W = S_{WR}$, called the **irreducible wetting-phase saturation**. At this saturation, the wetting fluid occupies a thin film adjacent to the solid grains. Reducing S_W below this value requires additional physics, such as evaporation of the wetting phase or a change in the interfacial tension between the two fluids N and W.

Imbibition starting at a saturation close to S_{WR} yields a different capillary pressure curve, called the **primary imbibition** curve. Here, the values of p_C are lower than the primary drainage values. In the absence of more complicated physics, as the imbibition progresses and S_W increases, the wetting fluid blocks more and more of the microscopic flow paths. Eventually, there are no connected microscopic flow paths available to the nonwetting fluid. As a result, it becomes impossible to reduce the nonwetting-phase saturation below $S_N = S_{NR}$, called the **irreducible nonwetting-phase saturation**. Thus, the primary imbibition curve has an endpoint at $S_W = 1 - S_{NR}$. Drainage starting at this saturation value yields another curve, called the **secondary drainage curve**. Saturation changes starting at values of S_W between S_{WR} and $1 - S_{NR}$ yield capillary pressure values lying along intermediate **scanning curves**.

Hysteresis implies that the model $p_C = p_C(S_W)$ cannot completely describe the effects of fluid-interface curvature on multifluid flows. Nevertheless, this functional relationship remains the most commonly used model, owing to its computational convenience and the difficulty of measuring hysteretic capillary pressure curves.

Petrophysicists associate certain attributes of rock-fluid systems with features of their capillary pressure curves. For example, because the drainage capillary pressure at a specific value of S_W reflects the pressure difference required for the nonwetting fluid to enter the narrowest pore spaces being invaded at that saturation, a primary drainage curve that is nearly flat over most of the saturation interval $(S_{WR}, 1 - S_{NR})$ indicates a relatively uniform distribution of pore-space diameters. A steeper primary drainage curve indicates a more heterogeneous mix of pore-space diameters.

6.2 Variably Saturated Flow

In 1931, before scientists had a widely accepted model for multifluid flows in porous media, American physicist Lorenzo A. Richards [129] proposed a model of two-fluid flow in soils, where the fluid phases of interest are water (W) and air (A). Richards based his work on a remarkable analysis in 1907 by another American physicist, Edgar Buckingham, [30] for the US Department of Agriculture.

6.2.1 Pressure Head and Moisture Content

The Richards model simplifies the physics of water flows in soils by assuming that the pressure in the air phase always equals the atmospheric pressure. This assumption effectively restricts the applicability of the model to near-surface settings, in which the nonwetting air phase is connected to the atmosphere. Specifically, we assume that $p_A = 0$, which is the **gauge pressure** of air at Earth's surface. This assumption eliminates the need to solve a separate flow equation for air, leaving only the flow equation for water. The assumption also implies that

$$p_C = p_A - p_W = -p_W > 0. \tag{6.3}$$

Figure 6.7 shows a typical distribution of water in soil near Earth's surface. As introduced in Section 4.3, in the **vadose zone**, $S_W < 1$, and, by Eq. (6.3), $p_W < 0$. The lower boundary of the vadose zone is the **water table**, defined as the depth where $p_W = 0$. Beneath the water table is the **phreatic zone**, where $S_W = 1$. In this zone, $p_W > 0$, increasing with depth. As mentioned in Section 4.3, immediately above the water table there is often a water-saturated capillary fringe, in which $p_W < 0$. Water added at Earth's surface, for example by rain, changes the boundary condition at the top of the soil column, and water subsequently moves downward through the soil. We call this type of flow **variably saturated**.

In parallel with groundwater hydrologists' preference for using piezometric head instead of pressure, soil scientists traditionally convert the negative water

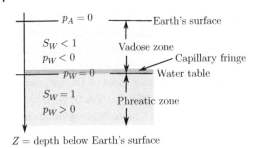

Figure 6.7 Schematic profile of soil in the near-surface region, showing the vadose zone ($p_W < 0$), the water table ($p_W = 0$), and the phreatic zone ($p_W > 0$).

pressure in the vadose zone to a function having dimension L, defining the **pressure head** or **tension head**,

$$\Psi = \frac{p_W}{\gamma g} = -\frac{p_C}{\gamma g}, \tag{6.4}$$

where γ denotes the true density of water. For variably saturated soils, $\Psi < 0$.

Soil scientists measure pressure head using the height of water in a **tensiometer**, a device consisting of a tube with a water-saturated porous cup at the soil interface, as illustrated schematically in Figure 6.8. The porous cup allows only water to flow into the tube from the soil, since there are no connected microscopic flow channels available to the air. The other end of the tube is open to the atmosphere. Water rises in the tube to a level given by $x_3 = \Psi - Z$, where $Z(\mathbf{x})$ is the depth of the point $\mathbf{x}$ below Earth's surface.

Soil scientists also traditionally refer to the **moisture content** $\Theta = \phi S_W$, which represents the volume of water per unit volume of the soil–air–water mixture. Introduction of this variable recasts the functional relationship $p_C(S_W)$ for capillary pressure as $\Psi(\Theta)$. In 1980, Dutch soil scientist Rien Van Genuchten [150] proposed a widely used closed-form expression commonly used to fit measured data for $\Psi(\Theta)$. The graph of $-\Psi(\Theta)$ resembles that of $p_C(S_W)$, as shown

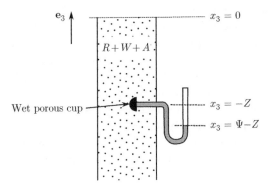

Figure 6.8 Schematic diagram of a tensiometer showing the water-saturated porous cup at depth Z. The cup is permeable only to water. The water level in the tube connected to the cup indicates the value of the pressure head Ψ.

Figure 6.9 Graph of a typical pressure head $\Psi(\Theta)$ for variably saturated flow in a soil. This graph neglects hysteresis.

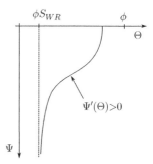

in Figure 6.9. Therefore, we expect $\Psi'(\Theta) > 0$. It is common to neglect the effects of hysteresis and to treat $\Psi(\Theta)$ as an invertible function, so that $\Theta = \Theta(\Psi)$, with $\Theta'(\Psi) > 0$.

6.2.2 The Richards Equation

The derivation of a flow equation for water in the vadose zone follows the usual motif: Substitute Darcy's law into the mass balance. We begin with the mass balance.

Exercise 6.6 *Assuming that there is no mass exchange between soil and water, show that, in terms of moisture content Θ, the mass balance for water in variably saturated flow takes the form*

$$\frac{\partial}{\partial t}(\gamma\Theta) + \nabla \cdot (\gamma\phi\mathbf{v}) = 0,$$

where $\mathbf{v}$ denotes the water velocity.

Since the water density is practically constant in this setting,

$$\frac{\partial\Theta}{\partial t} + \nabla \cdot (\phi\mathbf{v}) = 0. \tag{6.5}$$

Following Buckingham ([30]; see also [118]), Richards assumed that water flows according to a modified version of Darcy's law:

$$\phi\mathbf{v} = -K(\Theta)\left(\frac{1}{\gamma g}\nabla p_W - \nabla Z\right) = -K(\Theta)(\nabla\Psi + \mathbf{e}_3). \tag{6.6}$$

Here, K is the **unsaturated hydraulic conductivity**, having dimension LT^{-1}. The dependence of K on Θ reflects the increase in microscopic flow paths available to water, owing to the decrease in air saturation, as Θ increases. Figure 6.10 shows a typical shape for the graph of $K(\Theta)$, [65, 151], with $K \geqslant 0$.

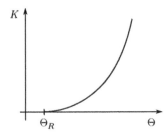

Figure 6.10 Graph of a typical unsaturated hydraulic conductivity $K(\Theta)$ for variably saturated flow in soil. The symbol Θ_R denotes the irreducible water content, corresponding to the irreducible wetting-phase saturation on the capillary pressure diagram shown in Figure 6.6.

Substituting for the filtration velocity $\phi\mathbf{v}$ in the mass balance equation (6.5) using Eq. (6.6) yields the **Richards equation**,

$$\boxed{\frac{\partial\Theta}{\partial t} - \nabla\cdot[K(\Theta)\nabla\Psi(\Theta) + K(\Theta)\mathbf{e}_3] = 0.} \tag{6.7}$$

For purely vertical flow, such as downward infiltration resulting from rainfall or irrigation, Eq. (6.7) reduces to

$$\frac{\partial\Theta}{\partial t} - \frac{\partial}{\partial x_3}\left[K(\Theta)\left(\frac{\partial\Psi}{\partial x_3}(\Theta) + 1\right)\right] = 0.$$

Exercise 6.7 *Show that, in terms of the depth $Z(\mathbf{x}) = -x_3$, Eq. (6.7) becomes*

$$\frac{\partial\Theta}{\partial t} - \frac{\partial}{\partial Z}\left[K(\Theta)\left(\frac{\partial\Psi}{\partial Z}(\Theta) - 1\right)\right].$$

In purely horizontal flow, with $x = x_1$, Eq. (6.7) reduces to the following nonlinear analog of the heat equation:

$$\frac{\partial\Theta}{\partial t} - \frac{\partial}{\partial x}\left[K(\Theta)\frac{\partial\Psi}{\partial x}(\Theta)\right] = 0. \tag{6.8}$$

6.2.3 Alternative Forms of the Richards Equation

Equation (6.7) is a **mixed formulation** of the Richards equation, since the time derivative operates on Θ while the spatial derivatives operate on Ψ. There are two ways to reformulate Eq. (6.7) in terms of a single principal unknown. The first is the **head-based formulation**: Write $\Theta = \Theta(\Psi)$ and $K = K(\Theta(\Psi))$, then use the chain rule to obtain

$$\Theta'(\Psi)\frac{\partial\Psi}{\partial t} - \nabla\cdot[K(\Theta(\Psi))(\nabla\Psi + \mathbf{e}_3)] = 0. \tag{6.9}$$

The positive coefficient $\Theta'(\Psi)$ is the **specific moisture capacity**.

The second reformulation is the **moisture-content formulation**. Here we use the chain rule to write $\nabla\Psi(\Theta) = \Psi'(\Theta)\nabla\Theta$, getting

$$\frac{\partial\Theta}{\partial t} - \nabla\cdot[D_W(\Theta)\nabla\Theta + K(\Theta)\mathbf{e}_3] = 0.$$

The nonnegative coefficient $D_W(\Theta) = K(\Theta)\Psi'(\Theta)$ is the **soil moisture diffusivity**, having dimension L^2T^{-1}. This formulation is especially useful in modeling purely horizontal flows, for which Eq. (6.8) becomes

$$\frac{\partial\Theta}{\partial t} - \frac{\partial}{\partial x}\left[D_W(\Theta)\frac{\partial\Theta}{\partial x}\right] = 0. \tag{6.10}$$

As appealing as either of these two reformulations may seem, neither may be a good choice for numerical work. To see how difficulties can arise, consider the discretization of Eq. (6.9) in time using a finite-difference approximation. In the head-based formulation, with Ψ as the principal unknown, we approximate the accumulation term using an expression of the following form:

$$\Theta'(\Psi)\frac{\partial\Psi}{\partial t} \simeq \Theta'(\Psi(\mathbf{x}, t^*))\frac{\Psi(\mathbf{x}, t + \Delta t) - \Psi(\mathbf{x}, t)}{\Delta t}. \tag{6.11}$$

Regardless of how one addresses other questions—such as how to approximate spatial derivatives or where, in time, to evaluate them—the approximation (6.11) confronts the numerical analyst with the unavoidable question of where, in time, to evaluate the nonlinear coefficient $\Theta'(\Psi(\mathbf{x}, t^*))$. Because no easily computed value of t^* correctly represents the time interval $[t, t + \Delta t]$ throughout the spatial domain, it is highly unlikely that the numerical solution will conserve mass globally. For a partial differential equation (PDE) derived from the mass balance, this shortcoming is significant.

To conserve mass numerically, it is usually better to leave the flow equation in its original, mixed formulation (6.7). Using this formulation, one can discretize in time using an approximation of the form

$$\frac{\partial\Theta}{\partial t}(\mathbf{x}, t) \simeq \frac{1}{\Delta t}\left[\Theta(\Psi(\mathbf{x}, t + \Delta t)) - \Theta(\Psi(x, t))\right],$$

using, for example, Newton's method to calculate successively better approximations of $\Theta(\Psi(\mathbf{x}, t + \Delta t))$ at every time step. For numerical methods based on this idea, see [6, 32].

6.2.4 Wetting Fronts

One additional aspect of variably saturated flows deserves comment, namely the effects of nonlinearity on qualitative properties of the solutions Θ. Equation (6.10) is a nonlinear analog of the heat equation. Figures 6.9 and 6.10 suggest that the nonlinear soil moisture diffusivity $D_W(\Theta)$ behaves approximately like $(\Theta - \Theta_R)^n$ for some exponent $n > 1$, a form that recalls the porous medium equation analyzed in Section 4.3. In particular, the parabolic nature of the PDE degenerates at the endpoint moisture content Θ_R.

As with the porous medium equation, this degeneracy allows sharp **wetting fronts** to propagate at finite speeds through initially dry soil, as depicted in

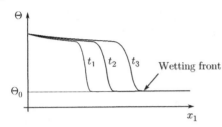

Figure 6.11 An advancing moisture content front showing a wetting front, downstream of which the moisture content remains undisturbed at its initial, dry-soil value Θ_0.

Figure 6.11. The distinctive feature of a wetting front is that, as it propagates, there remains a downstream region in which the water content remains at its irreducible value Θ_R. This behavior stands in contrast to solutions of the linear heat equation, which have infinite propagation speed (with rapid decay). In the linear case, a disturbance instantaneously affects solution values throughout the domain, as described in Section 4.2 in connection with the Theis model. For a deeper mathematical discussion of wetting fronts and their implications for numerical models, see [61].

6.3 Two-fluid Flows

6.3.1 The Muskat–Meres Model

In 1936, Russian-born American engineer Morris Muskat and American geophysicist Milan W. Meres [106]—citing experimental work by R.D. Wyckoff and H.G. Botset [159]—proposed modeling the flow of immiscible fluid phases through a porous medium by extending Darcy's law to all of the fluids. The **Muskat–Meres model** takes the following form:

$$\phi_\alpha \mathbf{v}_\alpha = -\frac{k_\alpha}{\mu_\alpha}(\nabla p_\alpha - \gamma_\alpha g \nabla Z), \tag{6.12}$$

where the index α ranges over all fluid phases. Here, as with the single-fluid version of Darcy's law, the filtration velocity $\phi_\alpha \mathbf{v}_\alpha$ represents the apparent velocity that one obtains by dividing the volumetric flow rate ($L^3 T^{-1}$) of fluid phase α across a surface by the total surface area.

The key difference between this model and the single-fluid version of Darcy's law is the use of an **effective permeability** k_α for each fluid phase, instead of the rock permeability k discussed in Section 3.7. Analogous to the unsaturated hydraulic conductivity introduced earlier by Buckingham [30], the factor k_α is no longer strictly a rock property, since it is supposed to account for the interference to the flow of fluid phase α arising from the presence of the other fluid. We expect $0 \leqslant k_\alpha \leqslant k$, the precise value depending on the fluid-phase saturations. If only two fluids N and W are present, then only one of the saturations is independent, since $S_N = 1 - S_W$.

It is common to decompose the effective permeability as $k_\alpha = k k_{r\alpha}$, where the **relative permeability** $k_{r\alpha}$ accounts for all of the effects associated with fluid–fluid resistance for the particular rock being analyzed. Paralleling the modeling assumptions for capillary pressure discussed in Section 6.1, the most common model for relative permeabilities assumes that they are functions of fluid saturation: For a two-fluid system, $k_{r\alpha} = k_{r\alpha}(S_W)$.

At the current state-of-the-art, one cannot determine the relative permeabilities on theoretical grounds alone. They must be measured for each system of fluids and rock. Typical two-fluid relative permeability curves have the shapes shown in Figure 6.12 and exhibit the following features:

1. They obey the inequalities $0 \leqslant k_{r\alpha} < 1$ and $k_{rN} + k_{rW} < 1$.
2. The wetting-fluid relative permeability $k_{rW}(S_W) = 0$ for $0 \leqslant S \leqslant S_{WR}$, and $k_{rN}(S_W) = 0$ for $1 - S_{NR} \leqslant S_W \leqslant 1$. Here S_{WR} and $1 - S_{NR}$ are the endpoint saturations seen in the capillary pressure curves for the fluid–fluid–rock system.
3. Both k_{rN} and k_{rW} are concave up, that is, $k_{rN}''(S_W)$ and $k_{rW}''(S_W)$ are positive for $S_{WR} < S_W < 1 - S_{NR}$.

Not shown in Figure 6.12—and often neglected in numerical models—is the experimental observation that relative permeabilities, like capillary pressures, exhibit hysteresis. This fact shows that the commonly used models for relative permeability fail to reflect some important physics.

As with capillary pressure curves, petrophysicists infer a variety of properties of the rock–fluid system from features of the relative permeability curves. One example is wettability. At the microscopic scale, at its residual saturation S_{NR}, the nonwetting fluid tends to form isolated blobs occupying the middle of the pore spaces. This configuration hinders the flow of the wetting fluid. The blocking is more effective than that imposed by the wetting fluid, which, at its residual saturation S_{WR}, tends to occupy a thin film adjacent to the rock grains and the concave

Figure 6.12 Typical relative permeability curves.

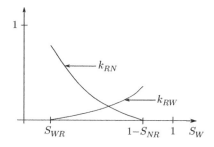

crevices between them. Based on this reasoning, we expect the endpoint relative permeability $k_{RN}(S_{WR})$ of the nonwetting fluid to be greater than the endpoint relative permeability $k_{RW}(1 - S_{NR})$ of the wetting fluid, as shown in Figure 6.12. It is common to use the ratio $k_{RN}(S_{WR})/k_{RW}(1 - S_{NR})$ as an indicator of wettability: The larger the ratio, the more strongly the wetting fluid W wets the rock in the presence of the nonwetting fluid N.

Because k_{rN} and k_{rW} both depend on S_W, Eq. (6.12) are coupled. They are also nonlinear, since the fluid saturations are unknowns to be determined as part of the solution process, as subsequent sections discuss in more detail. The coupled, nonlinear nature of Eq. (6.12) and their extensions to larger numbers of fluid phases have spawned a wide variety of analytical and numerical solution techniques, some of which we touch upon in the remainder of this section.

6.3.2 Two-fluid Flow Equations

As usual, we obtain flow equations by substituting the extension (6.12) of Darcy's law for filtration velocities into the mass balance equations. For a two-fluid flow system, the mass balance equations are

$$\frac{\partial}{\partial t}(\phi_\alpha \gamma_\alpha) + \nabla \cdot (\phi_\alpha \gamma_\alpha \mathbf{v}_\alpha) = r_\alpha, \quad \alpha = N, W.$$

In the absence of interphase mass transfer, $r_N = r_W = 0$, and substituting Eq. (6.12) for $\alpha = N, W$ yields the **two-fluid flow equations**:

$$\boxed{\frac{\partial}{\partial t}(\phi S_\alpha \gamma_\alpha) = \nabla \cdot \left[\frac{\gamma_\alpha k k_{r\alpha}}{\mu_\alpha}\left(\nabla p_\alpha - \gamma_\alpha g \nabla Z\right)\right], \qquad \alpha = N, W,} \qquad (6.13)$$

since $\phi_\alpha = \phi S_\alpha$.

Equations (6.13) provide two coupled, nonlinear PDEs. If we assume that the functions $\phi, \gamma_N, \gamma_W, k, k_{rN}, k_{rW}, \mu_N, \mu_W$ are either known or computable once we know the fluid pressures and saturations, we still have four principal unknowns: p_N, p_W, S_N, and S_W. Mathematical closure of the system therefore requires two additional equations. For these, we use the capillary pressure relationship

$$p_N = p_W + p_C(S_W), \qquad (6.14)$$

assuming that we have measured data defining $p_C(S_W)$, and the saturation restriction

$$S_N + S_W = 1 \qquad (6.15)$$

from Section 6.1. The nonlinearities now appear more explicitly: The functions $p_C(S_W), k_{rN}(S_W)$, and $k_{rW}(S_W)$ all depend on principal unknowns.

6.3.3 Classification of Simplified Flow Equations

Before trying to solve problems involving Eqs. (6.13)—(6.15), it is useful to classify the PDEs. For this task, we simplify the physics, at least temporarily. Assume for the moment that γ_N, γ_W, and ϕ are constant and that gravity effects are negligible, as in purely horizontal flows where $\nabla Z = 0$. Since $\partial S_N / \partial t = -\partial S_W / \partial t$, Eq. (6.13) reduce to

$$\phi \frac{\partial S_W}{\partial t} = \nabla \cdot (\lambda_W \nabla p_W),$$

$$-\phi \frac{\partial S_W}{\partial t} = \nabla \cdot (\lambda_N \nabla p_N), \tag{6.16}$$

where the symbols

$$\lambda_\alpha = \frac{k k_{r\alpha}}{\mu_\alpha}, \quad \alpha = N, W,$$

stand for the fluid **mobilities**.

Equations (6.16) provide two PDEs, which we may regard as governing the two principal unknowns p_N and p_W. The relationships (6.14) and (6.15), together with constitutive relationships for λ_N and λ_W, close the system. This view of the two-fluid flow equations motivates a class of numerical solution strategies based on **simultaneous solution (SS)** of the PDEs, first introduced in 1959 by American mathematicians Jim Douglas et al. [47].

Another orchestration leads to a different view and motivates an alternative numerical solution strategy. Adding Eqs. (6.16) gives

$$\nabla \cdot (\phi \mathbf{v}) = 0, \tag{6.17}$$

where

$$\phi \mathbf{v} = -\lambda_W \nabla p_W - \lambda_N \nabla p_N$$

denotes the total volumetric flow rate of fluid per unit area, having dimension LT^{-1}.

Exercise 6.8 *Show that Eq. (6.17) is equivalent to a **pressure equation**,*

$$\nabla \cdot (\lambda \nabla p) = \nabla \cdot \left(\frac{\lambda_W - \lambda_N}{2} \nabla p_C \right), \tag{6.18}$$

where

$$\lambda = \lambda_N + \lambda_W, \quad p = \frac{p_N + p_W}{2}$$

*denote the **total mobility** and **average pressure**, respectively.*

Provided the total mobility $\lambda(S_W)$ is positive and bounded away from zero, Eq. (6.18) furnishes an elliptic PDE for the average pressure p.

Exercise 6.9 *Show that the flow equation* (6.16) *for W is equivalent to the following* **saturation equation**:

$$\phi\frac{\partial S_W}{\partial t} = -\nabla \cdot [\phi f(S_W)\mathbf{v}] - \nabla \cdot [f(S_W)\lambda_N(S_W)p'_C(S_W)\nabla S_W], \tag{6.19}$$

where

$$f(S_W) = \frac{\lambda_W(S_W)}{\lambda(S_W)}. \tag{6.20}$$

We call the dimensionless function $f(S_W)$ defined in Eq. (6.20) the **fractional flow** function. It gives the fraction of *flowing* fluid that is phase W. Recall that S_W is the fraction of all fluid that is phase W. When the irreducible saturations S_{NR} and S_{WR} are not zero, some of the fluid is immobile.

Figure 6.13 shows the graph of a typical fractional flow function, consistent with the relative permeability curves shown in Figure 6.12. As Section 6.4 shows, the curve's S shape, with an inflection point at a saturation value lying between S_{WR} and $1 - S_{NR}$, has important implications for the solution of the flow equations.

Exercise 6.10 *Use the fact that* $p'_C(S_W) < 0$, *as shown in Figure* 6.6, *to show that Eq.* (6.19) *is parabolic at all saturation values for which*

$$f(S_W)\lambda_N(S_W)p'_C(S_W) \neq 0.$$

In oilfield applications, pressure gradients attributable to pumping at wells often dominate capillary pressure gradients as drivers of the flow. Neglecting the factor $\nabla p_C(S_W) = p'_C(S_W)\nabla S_W$ in Eq. (6.19) reduces it to a nonlinear, first-order hyperbolic PDE for the wetting-fluid saturation S_W:

$$\phi\frac{\partial S_W}{\partial t} + \nabla \cdot [\phi f(S_W)\mathbf{v}] = 0. \tag{6.21}$$

Section 6.4 examines solutions to a one-dimensional version of this equation.

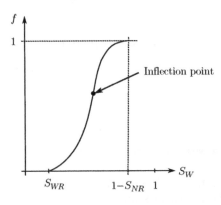

Figure 6.13 Typical fractional flow function associated with relative permeability curves having the shapes shown in Figure 6.12.

Equations (6.18) and (6.19) suggest a two-stage numerical solution strategy for the two-phase flow equations: First, solve the elliptic pressure equation for p using the most recently computed values of S_W. Then use the newly computed pressure p—from which we calculate $\phi\mathbf{v}$—to update S_W via the saturation equation. This orchestration, known as the **Implicit-Pressure-Explicit-Saturation (IMPES)** method, is commonly attributed to American engineers Herbert Stone and A.O. Garder [142]. For further details on the SS and IMPES methods, along with other numerical solution strategies for multifluid flow in porous media, see [11] and [35].

6.4 The Buckley–Leverett Problem

6.4.1 The Saturation Equation

In a seminal 1942 paper, American engineers Stuart E. Buckley and Miles C. Leverett [29] developed a simple model of two-fluid flow that sheds light on the mechanics of oil recovery by **immiscible displacement**, including gas injection and waterflooding. This model and the methods used to solve it have inspired many subsequent analyses of enhanced oil recovery methods and of nonlinear first-order PDEs more generally. In its simplest form, the Buckley–Leverett problem helps elucidate the hyperbolic nature of the saturation equation (6.21) and the shock-like fluid flows that drive oil through the rock toward production wells.

Several simplifying assumptions reduce the problem to a first-order hyperbolic PDE in one space dimension.

1. The flow of both the wetting and nonwetting fluids is incompressible and horizontal, as drawn in Figure 6.14. Thus $\nabla Z = \mathbf{0}$, and there is a coordinate system in which

$$\mathbf{v} = -\frac{1}{\phi}\left(\lambda_W \nabla p_W + \lambda_N \nabla p_N\right) = (v, 0, 0),$$

where v is a positive constant.
2. The porosity ϕ and fluid densities γ_N, γ_W are constant.
3. The term $\nabla p_C(S_W)$ is negligible, that is, the capillary effect is small, as a driver of the flow, compared with the effects of applied pressure gradients.

Figure 6.14 One-dimensional flow geometry used in the Buckley–Leverett problem.

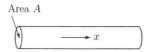

Exercise 6.11 *Assume that there is no leakage from the tube of constant cross-sectional area A illustrated in Figure 6.14. Integrate the saturation equation (6.21) over the cross section to get the one-dimensional, nonlinear equation*

$$\frac{\partial S}{\partial t} + v\frac{\partial f}{\partial x}(S) = 0. \tag{6.22}$$

Here and for the rest of this section, we simplify notation by writing $S_W = S$ and $x_1 = x$.

Equation (6.22) is the **Buckley–Leverett saturation equation**. Experience with first-order PDEs in Chapter 5 suggests the method of characteristics as a solution strategy for initial-value problems involving this equation. For concreteness, consider the following initial-value problem:

$$\frac{\partial S}{\partial t} + v\frac{\partial f}{\partial x}(S) = 0, \quad 0 < x < \infty, \quad t > 0, \tag{6.23}$$

$$S(x,0) = \begin{cases} 1 - S_{NR} - (x/L)(1 - S_{NR} - S_{WR}), & 0 \leqslant x < L, \\ S_{WR}, & L \leqslant x, \end{cases}$$

$$S(0,t) = 1 - S_{NR}, \quad t > 0.$$

These conditions prescribe values of the wetting-fluid saturation S along an initial curve in the (x, t)-plane consisting of the nonnegative x- and t-axes. Figure 6.15 shows the ramp-like graph of the initial saturation profile. Examining this continuous initial condition will show how, after finite time, the solution develops a discontinuity.

To implement the method of characteristics, apply the chain rule to Eq. (6.22) to get

$$\frac{\partial S}{\partial t} + vf'(S)\frac{\partial S}{\partial x} = 0. \tag{6.24}$$

The chain rule also gives the directional derivative of S along any continuously differentiable path $(x(s), t(s))$ in the (x, t)-plane:

$$t'(s)\frac{\partial S}{\partial t} + x'(s)\frac{\partial S}{\partial x} = \frac{dS}{ds}, \tag{6.25}$$

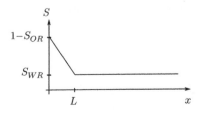

Figure 6.15 Ramp-shaped initial condition used in the initial-value problem (6.23).

at all points (x, t) where the wetting-fluid saturation $S(x, t)$ is differentiable. To find paths along which the left side of Eq. (6.24) coincides with the directional derivative on the left side of Eq. (6.25), set

$$t'(s) = 1, \quad x'(s) = vf'(S(x(s), t(s))). \tag{6.26}$$

These equations define the characteristic curves $(x(s), t(s))$, given by the ordinary differential equation

$$\frac{x'(s)}{t'(s)} = \frac{dx}{dt} = vf'(S). \tag{6.27}$$

The PDEs (6.24) and (6.25) imply that, along these curves, the characteristic equation

$$\frac{dS}{ds} = 0 \tag{6.28}$$

must hold.

As in the applications of the method of characteristics in Chapter 5, we conclude from Eq. (6.28) that S remains constant along characteristic curves. According to Eq. (6.27), these curves are lines, each of whose slope is the value of $vf'(S)$ at the point where the characteristic curve intersects the initial curve.

However, as in Section 5.4 this reasoning leads to nonsensical, multivalued solutions when characteristic curves associated with different saturation values cross. Figure 6.16a shows that this conflict arises for the S-shaped fractional flow function drawn in Figure 6.13: As S decreases from its maximum value $1 - S_{NR}$ to its minimum value S_{WR}, $vf'(S)$ first increases, then decreases, as shown in Figure 6.16b. Applied naïvely, the method of characteristics therefore eventually yields multivalued solutions, as shown in Figure 6.16c. Such solutions are physically impossible.

6.4.2 Welge Tangent Construction

The resolution to the apparent paradox of multivalued saturations parallels the discussion in Section 5.4. We allow the solution $S(x, t)$ to be a weak solution with a jump discontinuity, in the form of a saturation shock, at some moving spatial location $\Sigma(t)$. Figure 6.17 illustrates such a shock.

To select the correct location and strength of this discontinuity, we enforce an integral form of the mass balance law (6.22). Over any subinterval $[x_L, x_R]$ containing the locus $\Sigma(t)$,

$$\frac{d}{dt} \int_{x_L}^{x_R} S(x, t) \, dx = vf(S(x_L, t)) - vf(S(x_R, t)), \tag{6.29}$$

by the fundamental theorem of calculus. Equation (6.29) does not require $S(x, t)$ to be differentiable. The expression on the left side of Eq. (6.29) represents the rate of

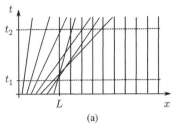

(a)

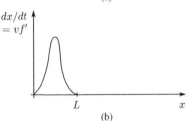

(b)

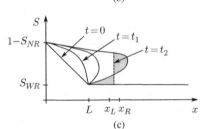

(c)

Figure 6.16 Graphic solution to the Buckley–Leverett problem: Part (a) shows the characteristic curves, which intersect because of the non-monotonic slope $vf'(S)$ plotted in (b). The diagram on (c) shows the graph of the solution $S(x, t)$ at several time levels, including the multivalued graphs predicted by the method of characteristics and the resolution enabled by the introduction of a saturation shock between x_L and x_R.

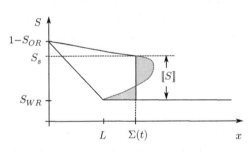

Figure 6.17 Saturation shock in the solution to the Buckley–Leverett problem.

accumulation of wetting fluid in the subinterval, while the difference appearing on the right represents the net flux of wetting fluid across the subinterval's endpoints.

Exercise 6.12 *Following the analysis leading to Eq. (5.54), show that the saturation shock $\Sigma(t)$ moves with speed $\Sigma'(t)$ given by the condition*

$$\Sigma'(t) = v\frac{[\![f(S)]\!]}{[\![S]\!]}. \tag{6.30}$$

Figure 6.18 Welge tangent construction to determine the saturation S_s at the upstream edge of the Buckley–Leverett saturation shock.

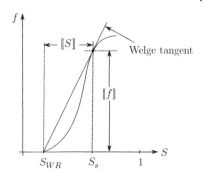

Equation (6.30) serves as a nondifferential version of the mass balance law, valid at points in the (x, t)-domain where the saturation S undergoes a jump discontinuity.

Figure 6.18 shows how to apply the condition (6.30) to the initial-boundary-value problem 6.23. Immediately downstream of the saturation discontinuity $\Sigma(t)$,

$$S(\Sigma+, t) = S_{WR},$$

since the saturation signal propagating from the left has yet to travel this far. Immediately upstream of $\Sigma(t)$,

$$S(\Sigma-, t) = S_s,$$

where the saturation value S_s satisfies the condition

$$vf'(S_s) = \frac{dx}{dt} = \frac{d\Sigma}{dt} = v\frac{[\![f]\!]}{[\![S]\!]}. \tag{6.31}$$

Graphically, this equation specifies S_s as the saturation value where a line passing through the initial value S_{WR} is tangent to the graph of $f(S)$. We call this line the **Welge tangent**, after American engineer Henry G. Welge, who published the construction in 1952 [154].

The construction illustrated in Figure 6.18 bears six remarks.

Remark 6.1 As shown in Figure 6.19, upstream of the saturation discontinuity the characteristic curves form a **rarefaction**—a term borrowed from gas dynamics. For the values of (x, t) in this region, the solution $S(x, t)$ varies smoothly.

Remark 6.2 Welge's choice of saturation discontinuity eliminates characteristic curve crossings. Of equal importance is the fact that it does so in a manner that connects every point (x, t) in the domain to the initial curve by a characteristic curve, as shown in Figure 6.19. Thus, the solution depends on the initial data. This dependence is a necessary condition for the problem to be well posed.

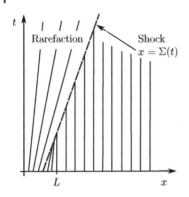

Figure 6.19 Characteristic curves for the Buckley–Leverett problem showing the rarefaction and the locus of the shock $\Sigma(t)$.

Remark 6.3 The Welge tangent construction resolves the problem of multivalued saturations by locating $\Sigma(t)$ so that the shaded areas to its left and right in Figure 6.16 are equal. Buckley, Leverett, and Welge associated this equal-area rule with mass balance–the same principle used above to derive Eq. (6.31). Welge credited this observation to the Hungarian-American mathematician John von Neumann.

Remark 6.4 The construction defines a weak solution to the Buckley–Leverett saturation equation. Although the concept of weak solutions has the virtue of admitting physically realistic solutions in cases where classical solutions may not exist, it arguably opens the door too wide. Weak solutions may not be unique, as Exercise 6.13 illustrates. The Welge tangent construction, which amounts to a graphic interpretation of the Rankine–Hugoniot condition, singles out the physically correct weak solution to the Buckley–Leverett saturation equation.

Remark 6.5 Physical considerations motivate an apparently independent criterion for identifying the correct weak solution. Recall that Eq. (6.22) neglects the capillary pressure gradient that appears in Eq. (6.19). Including this term yields the formally parabolic PDE

$$\frac{\partial S}{\partial t} + v \frac{\partial}{\partial x} f(S) - \frac{\partial}{\partial x}\left[D_{\text{cap}}(S)\frac{\partial S}{\partial x}\right] = 0,$$

where

$$D_{\text{cap}}(S) = -\frac{1}{\phi}f(S)\lambda_N(S)p'_C(S) \geqslant 0. \tag{6.32}$$

Equation (6.32) restores the diffusion-like effect needed to smooth the sharp front and yield a classical solution, as shown in Figure 6.20. This observation serves as a heuristic for the **vanishing viscosity principle** advanced by Russian mathematician Olga Oleinik [111] in the late 1950s: The correct solution is the limit,

Figure 6.20 Classical solution to the Buckley–Leverett problem with nonzero capillary pressure gradient.

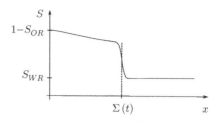

as $D_{\text{cap}} \to 0$, of solutions to the full equation (6.32). Hungarian-American mathematician Peter Lax [93] was among the first to recognize the relevance of limits of this type to numerical solution methods for PDEs.

Remark 6.6 The solution in Figure 6.17 exhibits a stability-like property for first-order PDEs that we employ again in Section 6.7, in connection with three-fluid flows: The saturation remains smooth only where the saturation speed is nonincreasing in the upstream direction. Otherwise, faster-moving saturations overtake slower-moving saturations, and a shock forms.

Exercise 6.13 *Sketch the weak solution associated with the characteristic curves shown in Figure 6.21.*

Exercise 6.14 *Sketch the solution $S(x, t)$ at several values of t for the following initial-value problem:*

$$\frac{\partial S}{\partial t} + v \frac{\partial f}{\partial x}(S) = 0, \quad 0 < x < \infty, \quad t > 0,$$

with initial condition

$$S(x, 0) = \begin{cases} S_s, & 0 \leqslant x < L, \\ S_{WR}, & L \leqslant x, \end{cases}$$

$$S(0, t) = S_s, \quad t > 0.$$

Here, the solution is discontinuous starting at $t = 0$, and the injected fluid saturation equals the shock saturation.

Figure 6.21 Characteristic curves for the Buckley–Leverett solution showing a shock associated with a physically incorrect weak solution.

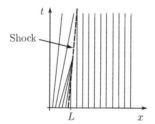

6.4.3 Conservation Form

The integrated form (6.29) of the Buckley–Leverett saturation equation prompts one additional remark. In general, a PDE of the form

$$\frac{\partial u}{\partial t} + \nabla \cdot \mathbf{F}(u) = 0, \tag{6.33}$$

where $\mathbf{F}$ is a vector-valued function, is a **conservation law**. We call $\mathbf{F}$ the **flux function** and say that Eq. (6.33) is in **conservation form**. In the saturation equation (6.21), $\mathbf{F}(u) = \phi f(u)\mathbf{v}$, and in the one-dimensional version (6.22), $F(u) = vf(u)$. In physical applications, first-order conservation laws commonly arise from balance laws such as the mass balance. By the chain rule, Eq. (6.33) is mathematically equivalent to the **nonconservation form**

$$\frac{\partial u}{\partial t} + \mathbf{F}'(u) \cdot \nabla u = 0.$$

For most numerical approximations, it is preferable to leave the PDE in conservation form. To see why, consider a typical Galerkin finite-element method. For Eq. (6.33), the method seeks an approximate solution $\hat{u}(\mathbf{x}, t)$ satisfying a system of equations having the form

$$\int_{\Omega} \left[\frac{\partial \hat{u}}{\partial t}(\mathbf{x}, t) + \nabla \cdot \mathbf{F}(\hat{u}(\mathbf{x}, t)) \right] \varphi_j(\mathbf{x}) \, dv = 0, \quad j = 1, 2, \dots, N. \tag{6.34}$$

Here, Ω denotes the spatial domain, and $\{\varphi_1, \varphi_2, \dots, \varphi_N\}$ is a set of test functions for which

$$\sum_{j=1}^{N} \varphi_j(\mathbf{x}) = 1, \quad \text{for all } \mathbf{x} \in \Omega.$$

Exercise 6.15 *Sum Eq. (6.34) and apply the divergence theorem to show that*

$$\frac{d}{dt} \int_{\Omega} \hat{u}(\mathbf{x}, t) \, dv = - \int_{\partial\Omega} \mathbf{F}(\hat{u}(\mathbf{x}, t)) \cdot \mathbf{n}(\mathbf{x}) \, ds. \tag{6.35}$$

Equation (6.35) asserts that the net flux of the approximate solution $\hat{u}(\mathbf{x}, t)$ across the boundary $\partial\Omega$ balances the rate of accumulation inside the domain Ω. In other words, when applied to the conservation form of the PDE, spatial discretization using the Galerkin finite-element method respects *a priori* the global conservation of mass that governs the true solution $u(\mathbf{x}, t)$.

6.4.4 Analysis of Oil Recovery

Welge [154] exploited the tangent construction to derive a simplified method for analyzing oil production in a fluid displacement project, such as a waterflood.

Consider the following initial-value problem for the Buckley–Leverett saturation equation:

$$\frac{\partial S}{\partial t} + v\frac{\partial f}{\partial x}(S) = 0, \quad 0 < x < \infty, \quad t > 0,$$

$$S(x, 0) = S_{WR}, \qquad x > 0,$$

$$S(0, t) = 1 - S_{NR}, \qquad t \geqslant 0.$$

This problem models a one-dimensional waterflood in which the water saturation at the inlet, $x = 0$, is the maximum possible value, $1 - S_{NR}$. The reservoir is initially saturated with oil, with water at its minimum possible saturation S_{WR}. A saturation shock with water saturation S_s, consistent with the Welge tangent construction, forms immediately.

By Eqs. (6.28) and (6.27), any water saturation value S propagates with speed

$$\frac{dx}{dt} = vf'(S).$$

Integrating this differential equation gives the spatial position of saturation value S at time t:

$$x(S) = vf'(S)\, t. \tag{6.36}$$

We now examine the arrival of the saturation shock, where the upstream saturation has value S_s, at a specified position x_B, which we regard as an outlet or oil production well. Petroleum engineers refer to this arrival as **breakthrough**; Eq. (6.36) shows that it occurs at time $t_B = x_B/[vf'(S_s)]$.

At time t_B, the flow rate of oil at the outlet undergoes an economically important change. For $0 < t < t_B$, the flow rate of oil at the outlet is high, since—under the model assumptions—the saturation shock effectively displaces all of the mobile oil—that is, oil in the saturation interval $1 - S_{NR} - S_{WR}$. However, at t_B the oil production rate drops sharply, and for $t > t_B$ the oil production rate tapers continuously toward zero, as illustrated in Figure 6.22. At some point, typically after breakthrough, the oil production rate drops below the threshold required to justify the cost of continued fluid injection.

Figure 6.22 Oil production rate as a function of time predicted by the Buckley–Leverett saturation equation. The production rate drops sharply at the breakthrough time t_B, when the saturation shock arrives at the outlet.

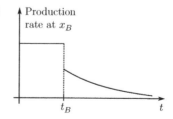

Based on this reasoning, Welge defined the theoretical **recovery efficiency** at breakthrough as follows:

$$E = \frac{(\text{Initial oil saturation}) - (\text{Average oil saturation at } t_B)}{1 - S_{NR} - S_{WR}}.$$

Here, the average oil saturation in the reservoir at time t_B is $1 - \overline{S}$, where

$$\overline{S} = \frac{1}{x_B} \int_0^{x_B} S(x, t_B) \, dx$$

is the average water saturation in the reservoir at t_B. Since the initial oil saturation is $1 - S_{WR}$,

$$E = \frac{\overline{S} - S_{WR}}{1 - S_{NR} - S_{WR}}. \tag{6.37}$$

Welge's simplified analysis reveals an elegant graphic interpretation for this expression.

Exercise 6.16 *Justify each step in the following calculation of the average water saturation:*

$$\overline{S} = \frac{1}{x_B} \left[S(x, t_B) x \Big|_0^{x_B} - \int_0^{x_B} x \frac{\partial S}{\partial x}(x, t_B) \, dx \right]$$

$$= S_s - \frac{1}{x_B} \int_{S(0, t_B)}^{S(x_B, t_B)} x(S) \, dS = S_s + \frac{1 - f(S_s)}{f'(S_s)}. \tag{6.38}$$

The right side of Eq. (6.38) is the value of S where the Welge tangent line has the value 1, as illustrated in Figure 6.23. This observation yields a graphically simple method for estimating the recovery efficiency at breakthrough. However, owing to flow phenomena that cannot be modeled in a one space dimension, this estimate serves, at best, as an optimistic upper bound on economic oil recovery. Section 6.5 explores this idea.

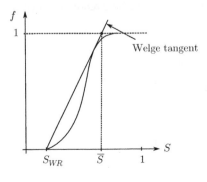

Figure 6.23 Welge's graphic construction of the average oil saturation $\overline{S}$ at the breakthrough time t_B.

6.5 Viscous Fingering

One-dimensional models of immiscible flow in porous media show how fluid displacement fronts, idealized as saturation shocks, can arise, but these models cannot capture all of the physics that bear on the design of underground processes. Of special significance in petroleum engineering is **viscous fingering**. This phenomenon results from frontal instabilities in the displacement of a resident fluid, such as oil, by a more mobile injected fluid, such as gas or water. This section treats the injected and displaced fluids as immiscible, denoting the fluid phases using the indices I and D, respectively.

In its simplest form, viscous fingering disrupts an initially regular displacement front, shown schematically in Figure 6.24a. The front becomes highly oscillatory, as depicted in Figure 6.24b, with fingers of injected fluid extending into the displaced fluid. Once established, these fingers of more mobile injected fluid grow longitudinally, creating high-mobility channels through which fluid I bypasses fluid D instead of displacing it. This frontal instability reduces the recovery efficiency far below its theoretical value, given by Eq. (6.37). Viscous fingering therefore has significant implications for the design of such oilfield processes as waterflooding, gas injection, and more costly enhanced oil recovery technologies.

British engineer S. Hill [72] identified the phenomenon in 1952 and provided an early mathematical analysis of front instability in vertical columns. Several years later, British mathematicians Philip G. Saffman and Sir Geoffrey I. Taylor [132] and Dutch engineers R.L. Chuoke et al. [36] independently published more complete analyses that have served as standards in a still-active literature on the subject; see [73, 117, 161]. The analysis presented here follows closely the reasoning developed in [132] and [36].

The idea is to examine the stability of an initially planar displacement front in the presence of a small perturbation in its shape. In natural porous media, such

Figure 6.24 Schematic diagram of viscous fingering: A more mobile injected fluid, having mobility λ_I, displaces a less mobile fluid having mobility λ_D. Part (a) shows the initially planar displacement front; part (b) shows the displacement front after unstable perturbations have grown.

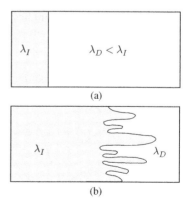

perturbations always arise, owing to microscopic heterogeneities. When the displacement is stable, the flow damps the perturbations, keeping the front nearly planar. However, when the displacement is unstable, small variations in the front grow into large fingers of injected fluid I. The stability analysis presented here does not reveal how the fingers grow once they arise. It simply indicates the conditions under which small perturbations grow or are suppressed.

6.5.1 The Displacement Front and Its Perturbation

To keep the analysis simple, adopt the following assumptions:

1. The porous medium is saturated with two immiscible fluids: an injected fluid I and a displaced fluid D.
2. The initial saturation of fluid D is $1 - S_{IR}$ throughout the medium.
3. Starting at time $t = 0$, we inject a mixture of fluids I and D, so that the saturation of fluid I upstream of the displacement front is S_s, the saturation at the upstream side of the Buckley–Leverett saturation shock. (See Section 6.4, especially Exercise 6.14.) This assumption, although strictly unrealistic from a process designer's perspective, simplifies the problem, since the fluid mobilities upstream of the displacement front remain constant.
4. The unperturbed displacement front Σ is planar, with a unit-length normal vector $\mathbf{n}$ that points into fluid D and lies at an angle θ to the vertical. In other words, if $\mathbf{g}$ denotes the gravitational acceleration vector, then $-\mathbf{n} \cdot \mathbf{g} = g \cos \theta$, as shown in Figure 6.25. Thus, $\theta = \pi/2$ when the displacement-front velocity is horizontal.
5. No fluid crosses the displacement front Σ.
6. The unperturbed displacement front Σ moves with constant, uniform velocity $\mathbf{v}_\Sigma = v_\Sigma \mathbf{n}$, where $v_\Sigma > 0$.
7. We adopt a rectangular coordinate system defined by the orthonormal basis $\{\mathbf{e}_1, \mathbf{e}_2, \mathbf{e}_3\}$, where $\mathbf{e}_3 = \mathbf{n}$, and $\mathbf{e}_1$ and $\mathbf{e}_2$ span the plane of the initial displacement front Σ. The origin lies on this plane at $t = 0$.
8. The fluid densities γ_I and γ_D and porosity ϕ are constant.

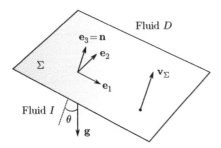

Figure 6.25 Geometry of the initially planar displacement front Σ separating the injected fluid I from the displaced fluid D.

Figure 6.26 Perturbation $\zeta(x_1, x_2, t)$ to the initial displacement front Σ.

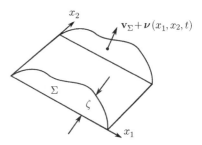

9. Capillary effects are negligible, so the displacement front Σ remains sharp, and the pressure is continuous across the displacement front.
10. By assumption 3, behind the fluid displacement front, fluid I has constant mobility $\lambda_I = kk_{rI}(S_s)/\mu_I$. Ahead of the interface, fluid D has constant mobility $\lambda_D = kk_{rD}(1 - S_{IR})/\mu_D$.

At $t = 0$, a perturbation distorts the displacement front Σ, in the x_3-direction, by a distance $\zeta(x_1, x_2, 0)$, as shown in Figure 6.26. The perturbed displacement front initially has velocity $\mathbf{v}_\Sigma + v(x_1, x_2, 0)$. We assume that, at least at early times, the function $\zeta(x_1, x_2, t)$ and its derivatives are small in magnitude and that the relative velocity $v(x_1, x_2, t)$ of the perturbed displacement front is also small in magnitude.

To work with a specific form of the perturbation, consider a typical Fourier component

$$\zeta(x_1, x_2, t) = \varepsilon(\boldsymbol{\omega}) \exp(\beta t + i\boldsymbol{\omega} \cdot \mathbf{x}),$$

having small initial amplitude $\varepsilon(\boldsymbol{\omega})$. Here, the real parameter β serves as a stability index for the displacement front: The perturbation grows exponentially if $\beta > 0$; it decays exponentially if $\beta < 0$. The nonzero propagation vector $\boldsymbol{\omega} = (\omega_1, \omega_2, 0)$ defines the frequencies with which the perturbation oscillates in the x_1- and x_2-directions. By the identity $\exp(i\omega_j x_j) = \cos(\omega_j x_j) + i\sin(\omega_j x_j)$, small values of $\|\boldsymbol{\omega}\|$ correspond to long wavelengths $2\pi/\|\boldsymbol{\omega}\|$. In keeping with the idea of a small initial perturbation, we treat the product $\|\boldsymbol{\omega}\|\|\zeta|$ as a small quantity.

Identifying the perturbed displacement front as a level surface of the function $F(x_1, x_2, x_3, t) = x_3 - \zeta(x_1, x_2, t)$, defined by the equation $F(\mathbf{x}, t) = 0$ as shown in Figure 6.27, helps in Section 6.5.2, where we determine the dynamics of the interface. Along any path $\mathbf{x}(t)$ for which $d\mathbf{x}/dt = v$, the value of F remains constant. Therefore, by the chain rule, along any such path,

$$
\begin{aligned}
0 = \frac{dF}{dt}(\mathbf{x}(t), t) &= \frac{d\mathbf{x}}{dt}(t) \cdot \nabla F(\mathbf{x}(t), t) + \frac{\partial F}{\partial t}(\mathbf{x}(t), t) \\
&= v \cdot \nabla F + \frac{\partial F}{\partial t} \\
&= -v_1 \frac{\partial \zeta}{\partial x_1} - v_2 \frac{\partial \zeta}{\partial x_2} + v_3 - \frac{\partial \zeta}{\partial t}.
\end{aligned}
\tag{6.39}
$$

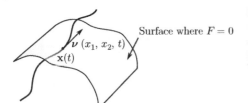

Figure 6.27 The perturbation as a level surface of the function $F(\mathbf{x}, t) = x_3 - \zeta(x_1, x_2, t)$, showing a path whose value $\mathbf{x}(t)$ always lies on the surface as t increases.

Since $\partial\zeta/\partial x_1$, $\partial\zeta/\partial x_2$, and the coordinates of v are all small in magnitude, we neglect products of these quantities. Doing so reduces Eq. (6.39) to

$$\frac{\partial\zeta}{\partial t}(x_1, x_2, t) = v_3(x_1, x_2, t). \tag{6.40}$$

6.5.2 Dynamics of the Displacement Front

We now examine how the perturbed displacement front moves by imposing the mass balance equations and Darcy's law for each fluid. Assumptions 6 ($\mathbf{v}_\Sigma$ is constant) and 8 (fluid densities remain constant) imply that the relative fluid velocities $v_\alpha = \mathbf{v}_\alpha - \mathbf{v}_\Sigma$ obey the following version of the mass balance:

$$\nabla \cdot v_\alpha = 0, \quad \alpha = I, D. \tag{6.41}$$

Darcy's law allows us to recast the problem of the displacement-front motion in terms of velocity potentials. Since λ_α and γ_α are constant upstream and downstream of the front, each fluid α has velocity

$$\begin{aligned}
\mathbf{v}_\alpha &= -\frac{\lambda_\alpha}{\phi}(\nabla p_\alpha - \gamma_\alpha g \nabla Z) \\
&= -\nabla\left[\frac{\lambda_\alpha}{\phi}(p_\alpha - \lambda_\alpha\gamma_\alpha g Z)\right] \\
&= -\nabla\left[\frac{\lambda_\alpha}{\phi}(p_\alpha + \lambda_\alpha\gamma_\alpha g\, x_3 \cos\theta)\right], \quad \alpha = I, D.
\end{aligned} \tag{6.42}$$

The last step follows from the observation that, in our coordinate system, the depth $Z(\mathbf{x}) = -x_3 \cos\theta$. It follows from Eq. (6.42) that the relative velocities $v_\alpha = \mathbf{v}_\alpha - v_\Sigma(0, 0, 1)$ are gradients of scalar potential fields:

$$v_\alpha = -\nabla\Phi_\alpha, \tag{6.43}$$

where

$$\Phi_\alpha = \frac{\lambda_\alpha}{\phi}p_\alpha - v_\Sigma x_3 + \frac{\lambda_\alpha}{\phi}\gamma_\alpha g\, x_3 \cos\theta, \quad \alpha = I, D. \tag{6.44}$$

Combining Eqs. (6.41) and (6.43) yields the Laplace equation for each of the velocity potentials Φ_I, Φ_D:

$$\begin{aligned}
\nabla^2\Phi_I &= 0, \quad x_3 < \zeta; \\
\nabla^2\Phi_D &= 0, \quad x_3 > \zeta.
\end{aligned} \tag{6.45}$$

To solve the PDEs (6.45) for Φ_I and Φ_D, we must impose boundary conditions. At the displacement front, the condition (6.40) holds. By the definition (6.43) of the velocity potentials and the assumption that no fluid crosses the displacement front, this condition yields

$$\frac{\partial \zeta}{\partial t} = \lim_{x_3 \to \zeta-} \mathbf{v}_I \cdot \mathbf{e}_3 = \lim_{x_3 \to \zeta+} \mathbf{v}_D \cdot \mathbf{e}_3,$$

that is,

$$\lim_{x_3 \to \zeta-} \frac{\partial \Phi_I}{\partial x_3} = \lim_{x_3 \to \zeta+} \frac{\partial \Phi_D}{\partial x_3}.$$

We call this condition, which constrains the fluid velocities in terms of the displacement-front velocity, a **kinematic boundary condition**.

For the remaining boundary conditions, we insist that the velocity perturbation die off as distance from the displacement front increases:

$$\lim_{x_3 \to -\infty} \Phi_I = 0; \quad \lim_{x_3 \to \infty} \Phi_D = 0.$$

Exercise 6.17 *Verify that the boundary-value problems for Φ_I and Φ_D have solutions*

$$\Phi_I = -\frac{\beta}{\|\boldsymbol{\omega}\|} \varepsilon(\boldsymbol{\omega}) \exp\left[\|\boldsymbol{\omega}\|(x_3 - \zeta) + \beta t + i\boldsymbol{\omega} \cdot \mathbf{x}\right]$$

$$= -\frac{\beta}{\|\boldsymbol{\omega}\|} \zeta \exp\left[\|\boldsymbol{\omega}\|(x_3 - \zeta)\right],$$

$$\Phi_D = \frac{\beta}{\|\boldsymbol{\omega}\|} \varepsilon(\boldsymbol{\omega}) \exp\left[-\|\boldsymbol{\omega}\|(x_3 - \zeta) + \beta t + i\boldsymbol{\omega} \cdot \mathbf{x}\right]$$

$$= \frac{\beta}{\|\boldsymbol{\omega}\|} \zeta \exp\left[-\|\boldsymbol{\omega}\|(x_3 - \zeta)\right]. \tag{6.46}$$

The functions in Eqs. (6.46) have the following values at the displacement front:

$$\lim_{x_3 \to \zeta-} \Phi_I = \frac{\beta}{\|\boldsymbol{\omega}\|} \zeta; \quad \lim_{x_3 \to \zeta+} \Phi_D = -\frac{\beta}{\|\boldsymbol{\omega}\|} \zeta. \tag{6.47}$$

6.5.3 Stability of the Displacement Front

To analyze the stability of the displacement front, we appeal to assumption 10 and impose the condition that the pressure is continuous across the front:

$$\lim_{x_3 \to \zeta-} p_I = \lim_{x_3 \to \zeta+} p_D. \tag{6.48}$$

Exercise 6.18 *Show that the relationship (6.48) holds if and only if, at $x_3 = \zeta$,*

$$M\Phi_D - \Phi_I - [(M-1)v_\Sigma + \lambda_I(\gamma_D - \gamma_I)g\cos\theta]\zeta = 0, \tag{6.49}$$

*where $M = \lambda_I/\lambda_D > 0$ is the **mobility ratio**.*

The identities (6.47) allow us to recast Eq. (6.49) in terms of the stability index β:

$$\beta = \frac{(M-1)v_\Sigma + \lambda_I(\gamma_D - \gamma_I)g\cos\theta]\|\omega\|}{M+1}.$$

Since $\|\omega\| > 0$, the displacement front is unstable ($\beta > 0$) if and only if

$$(M-1)v_\Sigma > \lambda_I(\gamma_I - \gamma_D)g\cos\theta. \tag{6.50}$$

Two special cases lend insight into design principles for immiscible displacements.

Case 6.1 *Horizontal flow.* Let $\theta = \pi/2$, that is, the flow is horizontal. In this case, condition (6.50) reduces to

$$M > 1.$$

Therefore, the displacement front is stable when the displacing fluid I is less mobile than the displaced fluid D. This observation serves as the main principle behind polymer flooding in oil reservoirs: Additives that increase the viscosity of injected water help keep its mobility smaller than that of the displaced oil.

Case 6.2 *Vertical flow.* When $\theta = 0$, that is, when the flow is upward, the instability condition (6.50) becomes

$$M > 1 + \frac{\lambda_I(\gamma_I - \gamma_D)}{v_\Sigma}.$$

In this case, provided the density ratio γ_I/γ_D of the two fluids is favorable–that is, $\gamma_I > \gamma_D$ –it is possible to stabilize the displacement even when $M > 1$, by controlling the injection rate and hence the speed v_Σ of the displacement front. A similar design principle holds for intermediate angles $\theta \in (0, \pi/2)$: When gravity is a factor, it is possible to stabilize the displacement of a less mobile fluid by a more mobile fluid by controlling the speed of the displacement front, provided the density ratio γ_I/γ_D is favorable.

Fingering phenomena also arise in flows in the vadose zone, where unfavorable density ratios result in gravity-driven instabilities in wetting fronts; see [122, 127, 135].

6.6 Three-fluid Flows

This chapter closes with two sections that introduce three-fluid flows in porous media. In this topic, even simple mathematical models entail many complexities and unsettled questions. The presentation here serves merely as a toehold on a large and active literature, prominent in which are [12, 22, 37, 44, 70, 87, 102, 109, 121].

Prototypically, three-fluid flows occur in petroleum reservoirs, where the three fluid phases are natural gas (G), oil (O), and water or formation brine (W). Often W is the most wetting phase, and G is the least wetting phase. Three-fluid flows also arise in groundwater contaminant hydrology, where the three fluids may be water, air, and a nonaqueous-phase liquid (NAPL), a setting first modeled numerically by American engineer Linda Abriola [1]. In these applications, too, the aqueous phase is often the most wetting phase, and the vapor phase—typically air contaminated with volatile organic compounds—is the least wetting phase.

In three-fluid flows through porous media, there are three fluid volume fractions, which we denote by ϕ_G, ϕ_O, and ϕ_W, and hence three saturations: $S_\alpha = \phi_\alpha / \phi$, for $\alpha = G, O, W$. Since

$$S_G + S_O + S_W = 1, \tag{6.51}$$

only two of the three saturations are independent. One can therefore specify any saturation state (S_G, S_O, S_W) as a point in, say, the (S_G, S_O)-plane, lying in the triangular region defined by the conditions $0 \leqslant S_G \leqslant 1$, $0 \leqslant S_O \leqslant 1$, and $S_G + S_O \leqslant 1$.

Equivalently, petroleum engineers commonly plot (S_G, S_O, S_W) as a point in a region bounded by an equilateral triangle. Figure 6.28 illustrates such a **ternary diagram**. The vertices labeled G, O, and W represent the points

$$(S_G, S_O, S_W) = \begin{cases} (1, 0, 0), \\ (0, 1, 0), \\ (0, 0, 1), \end{cases}$$

respectively. Any point in the ternary diagram corresponds to a saturation triple (S_G, S_O, S_W), represented by the point P in Figure 6.28 and given by the **method of intersections**. Draw line segments $\overline{GG'}$, $\overline{OO'}$, and $\overline{WW'}$ from each vertex of

Figure 6.28 A ternary diagram for three-phase saturations showing the geometric construction used to determine the saturations S_G, S_O, S_W of a point P in the triangle.

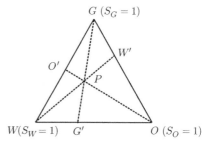

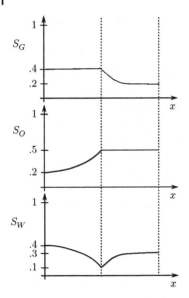

Figure 6.29 Saturation profiles for gas, oil, and water used in Exercise 6.19.

the triangle $\triangle GWO$, through the point P, to the opposite edge of the triangle, as shown in Figure 6.28. The saturations at P are the ratios

$$S_G = \frac{PG'}{GG'}, \quad S_O = \frac{PO'}{OO'}, \quad S_W = \frac{PW'}{WW'}. \tag{6.52}$$

Exercise 6.19 *Using Eqs. (6.52), sketch the path in the ternary diagram corresponding to the saturation profiles shown in Figure 6.29.*

6.6.1 Flow Equations

To determine the fluid saturations (S_G, S_O, S_W) at each point in the reservoir at each time t, we use the mass balance equations and the Muskat–Meres extension (6.12) of Darcy's law for each phase. The mass balances are

$$\frac{\partial}{\partial t}(\phi S_\alpha \gamma_\alpha) + \nabla \cdot (\phi S_\alpha \gamma_\alpha \mathbf{v}_\alpha) = r_\alpha, \quad \alpha = G, O, W. \tag{6.53}$$

For the remainder of this chapter, we assume that interphase mass transfer is negligible, so that $r_G = r_O = r_W = 0$. Chapter 7 relaxes this assumption.

The Muskat–Meres extension of Darcy's law gives the fluid-phase velocities as follows:

$$\phi S_\alpha \mathbf{v}_\alpha = -\frac{k k_{r\alpha}}{\mu_\alpha}(\nabla p_\alpha - \gamma_\alpha g \nabla Z), \quad \alpha = G, O, W. \tag{6.54}$$

Substituting Eq. (6.54) into Eq. (6.53) following the usual motif gives the **three-fluid flow equations:**

$$\frac{\partial}{\partial t}(\phi S_\alpha \gamma_\alpha) - \nabla \cdot \left[\frac{kk_{r\alpha}\gamma_\alpha}{\mu_\alpha}(\nabla p_\alpha - \gamma_\alpha g \nabla Z) \right] = 0, \quad \alpha = G, O, W.$$ (6.55)

Supplementing these PDEs are the saturation restriction (6.51); constitutive equations for $\phi, k, k_{r\alpha}$, and γ_α for $\alpha = G, O, W$; the geometric information required to define the depth $Z(\mathbf{x})$; and capillary pressure functions, discussed next.

6.6.2 Rock-fluid Properties

The rock-fluid properties $p_{C\alpha\beta}$ and $k_{r\alpha}$ merit several remarks. First, consider the capillary pressures $p_{C\alpha\beta}$:

$$p_{CGO}(S_G, S_O) = p_G - p_O,$$
$$p_{COW}(S_O, S_W) = p_O - p_W,$$
$$p_{CGW}(S_G, S_W) = p_G - p_W.$$ (6.56)

Only two of these relationships are independent, since $p_{CGW} = p_{CGO} + p_{COW}$.

Also, it is common to simplify the saturation dependencies shown in Eqs. (6.56). While in principle each capillary pressure is a function of two saturations, M.C. Leverett and W.R. Lewis [97] argued, based on experimental evidence, that the constitutive equations for p_{CGO} and p_{COW} can be approximated more simply as follows:

$$p_{CGO} = p_{CGO}(S_G), \qquad p_{COW} = p_{COW}(S_W).$$ (6.57)

This simplification remains in common use [11, p. 32], [35, p. 53]. Strictly speaking, none of the functional relationships shown in Eqs. (6.56) and (6.57) is correct, since three-phase capillary pressures exhibit hysteresis, just as in two-phase flows.

Three-phase relative permeabilities also deserve comment. Based on experiments in water-wet porous media, American engineer Arthur T. Corey et al. [40] suggested the non-hysteretic functional dependencies

$$k_{rG} = k_{rG}(S_G), \quad k_{rO} = k_{rO}(S_G, S_W), \quad k_{rW} = k_{rW}(S_W).$$ (6.58)

More recent experimental data in water-wet porous media suggest that these dependencies are realistic for k_{rO} and k_{rW} but that a relationship of the form $k_{rG} = k_{rG}(S_G, S_W)$ is more realistic for the gas phase; see [3] for a review.

Measuring three-phase relative permeabilities directly proves difficult and costly. For this reason, scientists and engineers have proposed a variety of methods for inferring three-phase relative permeability curves from laboratory measurements of two-phase systems. In the early 1970s, American engineer Herbert L. Stone [140, 141] pioneered two of the early methods for using two-phase

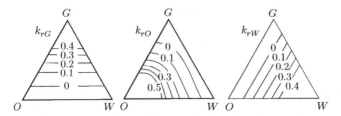

Figure 6.30 Hypothetical three-phase relative permeability contour plots showing the isoperms for $k_{rG}(S_G)$, $k_{rO}(S_G, S_W)$, and $k_{rW}(S_W)$.

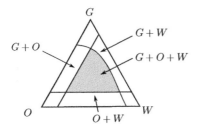

Figure 6.31 Saturation ternary diagram showing the reduced ternary diagram as a shaded region in which all three fluid phases are mobile. Outside the isoperms where $k_{r\alpha} = 0$ for $\alpha = G, O, W$, at most two fluid phases flow. In the diamond-shaped regions near the vertices of the diagram only one fluid phase flows.

flow data to calculate three-phase relative permeabilities. Since then numerous additional methods of this type have appeared; see [27, 48, 49, 121, 142] for examples.

Figure 6.30 illustrates three-phase relative permeabilities of the forms (6.58). Instead of trying to depict each function's graph as a surface above the saturation ternary diagram, the figure shows level sets of k_{rG}, k_{rO}, and k_{rG}, called **isoperms**.

Exercise 6.20 *Justify the depiction of isoperms for k_{rG} and k_{rW} as line segments in Figure* 6.30.

These isoperm plots make it clear that fully three-phase fluid flow occurs only inside a subregion of the saturation ternary diagram. We call this subregion the **reduced ternary diagram**. As Figure 6.31 shows, the loci of irreducible gas, oil, and water saturations form the boundaries of this subregion. At saturation values (S_G, S_O, S_W) lying outside these boundaries, at most two of the fluid phases are mobile, and in regions near the vertices of the ternary diagram only one fluid phase is mobile.

6.7 Three-fluid Fractional Flow Analysis

The system comprising Eqs. (6.55), (6.51), and (6.57), augmented by appropriate constitutive relationships, is daunting in complexity and, in general, solvable only via numerical approximations. Under certain simplifying assumptions, however, the system reduces to an equation set that is amenable to analysis by extending the methods used in Section 6.4 for the Buckley–Leverett problem [87, 132]. The resulting simplified model reveals an astonishing feature of the three-fluid flow problem.

6.7.1 A Simplified Three-fluid System

As in the Buckley–Leverett problem of Section 6.4, the first simplification is to restrict attention to one-dimensional flows in which gravitational effects play no role and capillary effects are negligible. These assumptions represent settings in which the flow is purely horizontal and in which the influence of applied pressure gradients, such as those imposed by pumping at wells, outweighs the influence of capillarity in driving the fluid flows. Under these assumptions, the three-fluid flow equations (6.55) reduce to the following form:

$$\frac{\partial}{\partial t}(\phi S_\alpha \gamma_\alpha) + \frac{\partial}{\partial x}(\phi S_\alpha \gamma_\alpha v_\alpha) = 0, \quad \alpha = G, O, W,$$

with one-dimensional filtration velocities

$$\phi S_\alpha v_\alpha = -\lambda_\alpha \frac{\partial p}{\partial x}.$$

Here, $\lambda_\alpha = k k_{r\alpha}/\mu_\alpha$ denotes the mobility of fluid phase α. Since capillarity is negligible in this model, all three fluids have the same pressure p.

Further simplification is possible under the special assumptions that the porosity ϕ and fluid densities γ_α are constant. Strictly speaking, the latter assumption is unrealistic for pressure-driven flows involving a gas phase, but it facilitates an analysis of the qualitative behavior of the system. Under these additional assumptions, the flow equations reduce even further to

$$\phi \frac{\partial S_\alpha}{\partial t} - \frac{\partial}{\partial x}\left(\lambda_\alpha \frac{\partial p}{\partial x}\right) = 0, \quad \alpha = G, O, W. \tag{6.59}$$

Exercise 6.21 *Following the logic of the IMPES formulation discussed in Section 6.3, add Eq. (6.59) over all fluid phases and use the saturation restriction (6.51) to obtain a* **pressure equation**,

$$-\frac{\partial}{\partial x}\left(\lambda \frac{\partial p}{\partial x}\right) = 0. \tag{6.60}$$

Here, $\lambda = \lambda_G + \lambda_O + \lambda_W$ stands for the total mobility. This tactic leaves two independent PDEs of the form (6.59).

The pressure equation (6.60) shows that the total filtration velocity

$$\phi v = -\lambda \frac{\partial p}{\partial x}$$

is a function at most of time. This result makes sense physically in a one-dimensional geometry in which all fluids flow incompressibly. The remainder of this section focuses on flows in which ϕv is a prescribed constant, determined by conditions at an inlet. Each fluid phase velocity is then $v_\alpha = f_\alpha v$, where, by analogy with Eq. (6.20), the ratio

$$f_\alpha(S_G, S_W) = \frac{\lambda_\alpha(S_G, S_W)}{\lambda(S_G, S_W)}$$

is the **fractional flow** function for phase α. This function gives the fraction of the flowing fluid that consists of phase α.

We are left with the two first-order saturation equations:

$$\frac{\partial S_G}{\partial t} + \frac{\partial F_G}{\partial x}(S_G, S_W) = 0,$$
$$\frac{\partial S_W}{\partial t} + \frac{\partial F_W}{\partial x}(S_G, S_W) = 0. \tag{6.61}$$

These equations, written in conservation form, involve the flux functions

$$F_G(S_G, S_W) = f_G(S_G, S_W)\, v, \quad F_W(S_G, S_W) = f_W(S_G, S_W)\, v.$$

Defining

$$\mathbf{S} = \begin{bmatrix} S_G \\ S_W \end{bmatrix}, \quad \mathbf{F(S)} = \begin{bmatrix} F_G(S_G, S_W) \\ F_W(S_G, S_W) \end{bmatrix},$$

reduces Eqs. (6.61) to a more compact vector form:

$$\boxed{\frac{\partial \mathbf{S}}{\partial t} + \frac{\partial}{\partial x}\mathbf{F(S)} = \mathbf{0}.} \tag{6.62}$$

The system (6.62) is a two-equation analog of the Buckley–Leverett saturation equation (6.22), with $\mathbf{F(S)}$ serving as the flux function.

6.7.2 Classification of the Three-fluid System

We now examine the classification of the system (6.62) of first-order PDEs. By the chain rule, the system is equivalent to the following equation:

$$\frac{\partial}{\partial t}\begin{bmatrix} S_G \\ S_W \end{bmatrix} + \begin{bmatrix} F_{GG} & F_{GW} \\ F_{WG} & F_{WW} \end{bmatrix} \frac{\partial}{\partial x}\begin{bmatrix} S_G \\ S_W \end{bmatrix} = \begin{bmatrix} 0 \\ 0 \end{bmatrix}. \tag{6.63}$$

Here, the derivative or Jacobian matrix of $\mathbf{F}$ has entries

$$F_{GG} = \frac{\partial F_G}{\partial S_G}, \quad F_{GW} = \frac{\partial F_G}{\partial S_W}, \quad F_{WG} = \frac{\partial F_W}{\partial S_G}, \quad F_{WW} = \frac{\partial F_W}{\partial S_W}.$$

In vector-matrix notation, the system (6.63) has the form

$$\frac{\partial \mathbf{S}}{\partial t} + \mathbf{F}'(\mathbf{S})\frac{\partial \mathbf{S}}{\partial x} = \mathbf{0}, \tag{6.64}$$

where $\mathbf{F}'(\mathbf{S})$ stands for the derivative of the flux function.

The analysis seeks self-similar solutions, as introduced in Section 4.2.4.

Exercise 6.22 *Show that the system (6.64) is invariant under the stretching transformation*

$$\xi = \varepsilon x, \quad \tau = \varepsilon t, \quad \eta = \varepsilon^0 \mathbf{S}.$$

It follows (see Eq. (4.35)) that self-similar solutions have the form

$$\mathbf{S}(x, t) = \mathbf{U}(\zeta), \quad \zeta = \frac{x}{t}.$$

Exercise 6.23 *Show that the system (6.64) reduces to the following system of ordinary differential equations for* $\mathbf{U}(\zeta)$:

$$\mathbf{F}'(\mathbf{U}(\zeta))\mathbf{U}'(\zeta) = \zeta\mathbf{U}'(\zeta). \tag{6.65}$$

One solution to Eq. (6.65) is $\mathbf{U}'(\zeta) = \mathbf{0}$, which implies that $\mathbf{U}(\zeta)$—and hence the saturation vector $\mathbf{S}(x, t)$—is constant. Of greater interest are nonconstant solutions. For any such solution $\mathbf{U}(\zeta)$, Eq. (6.65) requires that its tangent vector $\mathbf{U}'(\zeta)$ be an eigenvector of the Jacobian matrix $\mathbf{F}'(\mathbf{U}(\zeta))$, with eigenvalue ζ, for all values of the similarity variable ζ where $\mathbf{U}$ is differentiable.

Exercise 6.24 *Show that the eigenvalues of* $\mathbf{F}'$ *are roots of the quadratic equation*

$$\zeta^2 - (F_{GG} + F_{WW})\zeta + \det \begin{bmatrix} F_{GG} & F_{GW} \\ F_{WG} & F_{WW} \end{bmatrix} = 0, \tag{6.66}$$

where det *stands for the determinant.*

Hence at any value of the saturation vector $\mathbf{S}$, the derivative $\mathbf{F}'$ possesses two eigenvalues $\zeta_+(\mathbf{S})$ and $\zeta_-(\mathbf{S})$, given by

$$\zeta_\pm = \frac{1}{2}(F_{GG} + F_{WW})$$

$$\pm \frac{1}{2}\sqrt{(F_{GG} + F_{WW})^2 - 4(F_{GG}F_{WW} - F_{GW}F_{WG})}.$$

Exercise 6.25 *Show that the eigenvectors of* $\mathbf{F}'(\mathbf{S})$ *corresponding to the eigenvalues* $\zeta_\pm$ *have the forms*

$$\mathbf{U}'_\pm = \begin{bmatrix} U'_G \\ U'_W \end{bmatrix}_\pm, \tag{6.67}$$

where

$$\frac{U'_G}{U'_W} = \frac{\zeta_{\pm}(S) - F_{WW}(S)}{F_{WG}(S)} = \frac{F_{GW}(S)}{\zeta_{\pm}(S) - F_{GG}(S)}. \tag{6.68}$$

(Since an eigenvector is unique only up to a nonzero multiplicative constant, it suffices to specify the ratio of the two entries.)

We now confront the fact that solutions $\zeta_{\pm}(S)$ to a quadratic equation of the form (6.66) can conceivably be complex-valued for certain points S in the reduced ternary diagram. The nature of these solutions determines the classification of the system (6.61).

- If $\zeta_{\pm}(S)$ are both real at the saturation value S and the matrix $F'(S)$ possesses linearly independent eigenvectors of the form (6.67) (so the matrix is diagonalizable), then the system (6.64) is **hyperbolic** at S. An important special case arises when $\zeta_+(S) \neq \zeta_-(S)$, in which case the system is **strictly hyperbolic** at S.
- If $\zeta_+(S)$ and $\zeta_-(S)$ are both real and equal to each other but the eigenvectors (6.67) fail to be linearly independent, then the system is **parabolic** at S.
- In the remaining case, $\zeta_+(S)$ and $\zeta_-(S)$ are not real-valued but are complex conjugates of each other. In this case, the system is **elliptic** at S.

Conceivably, the classification of the system (6.64) may be different at different points S in the saturation ternary diagram.

6.7.3 Saturation Velocities and Saturation Paths

When the system (6.64) is strictly hyperbolic at all points in the reduced ternary diagram, we associate with each point S two distinct, real eigenvalues $\zeta_{\pm}$ and their corresponding, linearly independent eigenvectors $U'_{\pm}$. To interpret the eigenvalues, consider their defining equation,

$$F'U'_{\pm} = \zeta_{\pm}U'_{\pm}. \tag{6.69}$$

By the chain rule,

$$\frac{\partial U}{\partial x}(\zeta(x,t)) = U'(\zeta)\frac{\partial \zeta}{\partial x} = \frac{1}{t}U'(\zeta),$$

and hence Eq. (6.69) and the original system (6.63) yield

$$-\frac{\partial U}{\partial t} = F'\frac{\partial U}{\partial x} = \frac{1}{t}F'U'(\zeta) = \frac{1}{t}\zeta U' = \zeta_{\pm}\frac{\partial U}{\partial x}.$$

From previous applications of the method of characteristics (see Sections 5.4 and 6.4), we know that the resulting system

$$\frac{\partial U}{\partial t} + \zeta_{\pm}\frac{\partial U}{\partial x} = 0$$

of PDEs implies that $\mathbf{U}$ (hence $\mathbf{S}$) remains constant along the characteristic curves defined by the differential equations

$$\frac{dx}{dt} = \zeta_{\pm}.$$

Thus, passing through each point in the (x, t)-domain are two curves: a **fast** characteristic curve, obeying the differential equation

$$\frac{dx}{dt} = \zeta_{+}(\mathbf{S}(x, t)), \tag{6.70}$$

and a **slow** characteristic curve, obeying

$$\frac{dx}{dt} = \zeta_{-}(\mathbf{S}(x, t)). \tag{6.71}$$

Because any saturation value $\mathbf{S}$ travels with one of these velocities, we refer to the eigenvalues $\zeta_{\pm}$ as **saturation velocities**.

To interpret the eigenvectors, regard each value of $\mathbf{U}'_{+}$ as a vector associated with the corresponding point $\mathbf{S}$ in the reduced ternary diagram. The resulting vector field defines a family of **integral curves**—curves everywhere tangent to $\mathbf{U}'_{+}$—in $\mathbf{S}$-space and associated with saturation velocity ζ_{+}, as illustrated using solid curves in Figure 6.32. The vector field $\mathbf{U}'_{-}$ defines a different family of integral curves, associated with saturation velocity ζ_{-}, shown in Figure 6.32 as dotted curves.

Exercise 6.26 *Using Eq. (6.68), show that*

$$\frac{dS_G}{dS_W} = \frac{\zeta_{\pm}(\mathbf{S}) - F_{WW}(\mathbf{S})}{F_{WG}(\mathbf{S})}, \text{ provided } F_{WG}(\mathbf{S}) \neq 0;$$

$$\frac{dS_W}{dS_G} = \frac{\zeta_{\pm}(\mathbf{S}) - F_{GG}(\mathbf{S})}{F_{GW}(\mathbf{S})}, \text{ provided } F_{GW}(\mathbf{S}) \neq 0.$$

Solutions to the differential equations in Exercise 6.26 have the form $S_G = S_G(S_W)$ and $S_W = S_W(S_G)$, respectively. Hence, any change in the saturation value $\mathbf{S}$ proceeds along an integral curve in the reduced ternary diagram. Following [71], we call these curves **saturation paths**.

Figure 6.32 Two families of integral curves in the reduced ternary diagram defined by the eigenvectors $\mathbf{U}'_{\pm}$ of the flux matrix for three-phase flow. Arrowheads show the directions of increasing saturation velocity. The solid curves are associated with the fast saturation velocity ζ_{+}; the dotted curves are associated with the slow saturation velocity ζ_{-}.

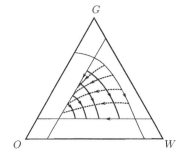

The saturation velocities and saturation paths play important roles in the solution of flow problems, as an example in Section 6.7.4 demonstrates. Following Remark 6.6 in Section 6.4.2, it is also important to ascertain whether the saturation velocities $\zeta_\pm$ increase or decrease along the respective saturation paths. Figure 6.32 shows a simple case, when the saturation velocity is monotonic along each saturation path.

6.7.4 An Example of Three-fluid Displacement

We now apply the ideas developed so far for three-fluid flows to analyze a simplified, hypothetical three-fluid displacement in a one-dimensional oil reservoir. For this purpose, assume that the system (6.63) is strictly hyperbolic at every point $\mathbf{S}$ in the reduced ternary diagram. This assumption bears discussion at the end.

The example involves a **Riemann problem**, in which we seek a self-similar solution to an initial-value problem having piecewise constant initial data on the x-axis:

$$\frac{\partial \mathbf{S}}{\partial t} + \mathbf{F}'(\mathbf{S})\frac{\partial \mathbf{S}}{\partial x} = \mathbf{0}, -\infty < x < \infty,$$

$$\mathbf{S}(x, 0) = \begin{cases} \mathbf{S}_L, & \text{if} -\infty < x \leqslant 0, \\ \mathbf{S}_R, & \text{if } 0 < x < \infty. \end{cases} \tag{6.72}$$

Assume that $\mathbf{S}_L \neq \mathbf{S}_R$, so the initial condition consists of a constant left saturation state $\mathbf{S}_L$, modeling the injected fluids, and a constant right saturation state $\mathbf{S}_R$, modeling the initially resident reservoir fluids. Assume also that only water is mobile at the injected saturation state $\mathbf{S}_L$ and that the initial saturation state $\mathbf{S}_R$ is a constant, three-fluid mixture in which only oil and gas are mobile, as illustrated in Figure 6.33.

The solution of Riemann problems for systems of first-order hyperbolic equations involves many subtleties that we do not fully explore here; for more details see [12, 81, 88]. Still, one can deduce several features of the solution to the problem (6.72) from observations made in the previous subsection. According to the results of Exercise 6.26, saturation values follow saturation paths. Figure 6.33 shows two saturation routes connecting $\mathbf{S}_L$ and $\mathbf{S}_R$ that follow such paths. Both are consistent with the method of characteristics, but only one is physically correct.

To determine which, we apply the stability-like condition observed in the two-fluid Buckley–Leverett problem of Section 6.4: For saturations to remain smoothly varying, saturation velocities must be nonincreasing as we move upstream from $\mathbf{S}_R$ to $\mathbf{S}_L$. Otherwise, faster-moving saturations overtake slower-moving ones, and a shock forms. Route 2 in Figure 6.33 cannot satisfy this condition, since the segment that follows a saturation path associated with a fast saturation velocity is upstream of the segment that follows a saturation path associated with a slow saturation velocity.

Figure 6.33 Two saturation routes connecting the constant states $\mathbf{S}_L$ and $\mathbf{S}_R$ via saturation paths in the Riemann problem (6.72). Arrows on the saturation paths indicate the directions of increasing saturation velocity.

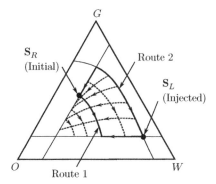

Figure 6.34 Saturation profile for the three-phase displacement modeled by the Riemann problem (6.72).

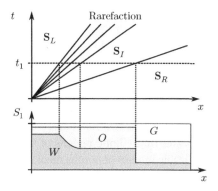

In contrast, on route 1 the upstream segment follows a saturation path associated with a slow saturation velocity, while the downstream segment follows a saturation path associated with a fast velocity. In addition, as the arrows in Figure 6.33 indicate, the upstream segment obeys the stability-like condition, since the saturation velocity increases as we move downstream. The saturation state $\mathbf{S}$ therefore varies smoothly along this segment, with values remaining constant along slow characteristic curves, as illustrated in Figure 6.34. Borrowing from the gas dynamics literature, we call this smoothly varying part of the solution a **rarefaction**.

On the downstream segment of route 1—the segment closer to $\mathbf{S}_R$—the saturation velocity increases in the upstream direction. Thus, upstream saturation values overtake downstream values, and a shock forms, as shown in Figure 6.34.

As Figure 6.34 shows, for a fixed time $t_1 > 0$, the saturation profile corresponding to saturation route 1 connects a self-similar, smoothly varying rarefaction to a saturation shock via a constant intermediate saturation state $\mathbf{S}_I$. The figure shows an oil bank moving downstream of the saturation shock, as expected in many oil recovery projects involving fluid injection.

The simplicity of this example belies the complexity of the three-phase Riemann problem. More complicated combinations of shocks and rarefactions arise, for example, when saturation velocities are not monotonic along saturation paths; see, for example, [12, 70].

The simplified three-fluid flow model admits even more exotic behavior. In 1986, American mathematicians John Bell, John Trangenstein, and Gregory Shubin [20] identified a class of three-fluid relative permeabilities, based on the Stone model (see Section 6.6), for which the system (6.64) of PDEs is not strictly hyperbolic everywhere in the saturation ternary diagram. In particular, they found regions in saturation space in which the system is elliptic. This finding is both astonishing and problematic. Well-posed elliptic problems require boundary conditions, and in the (x, t)-plane this requirement calls for the imposition of future values—that is, prescribed values at times later than those imposed on an initial curve, as illustrated in Figure 5.5b.

Since the discovery of elliptic regions in idealized three-fluid flows, a rich literature has grown around the topic, with varying interpretations; see, for example, [91, 143]. One interpretation [83, 84], not universally accepted, proposes treating strict hyperbolicity as a requirement for a realistic model of the form (6.63). This requirement imposes restrictions on properties of three-phase relative permeabilities beyond the assumptions inherent in the Stone model. On the other hand, numerical evidence from models based on capillary tube bundles (see Section 6.1) indicates that elliptic regions of saturation space can indeed arise in physically realizable models of porous media [78].

7

Flows With Mass Exchange

This last chapter examines multifluid flows in which the fluid phases exchange mass. Although interphase mass transfer occurs to some extent in all multifluid flows in porous media, in some applications its effects are too important to neglect.

Many enhanced oil recovery technologies fall into this category. For example, injecting carbon dioxide into oil reservoirs leads to the dissolution of carbon dioxide in the oil and to some evaporation of the lighter hydrocarbon species in the oil [91, Chapter 7]. As the process evolves, the oil phase becomes less viscous, and the interfacial tension between the oil and the injected fluid decreases. Under favorable conditions, this process develops into a highly efficient **miscible displacement**, in which very little oil remains in the reservoir zones swept by the injected fluid. Interphase mass transfer is also important in many groundwater contamination problems, for example when an immiscible nonaqueous-phase liquid (NAPL) invades an aquifer and volatile organic compounds dissolve in the aqueous phase [1].

Modeling processes of this type requires that we account not only for the physics of multifluid flows in porous media but also for changes in fluid properties that occur as chemical species cross the phase boundaries through evaporation and dissolution. We call models of such flows **compositional models**.

Because compositional models are so computationally intensive, this chapter provides only a brief overview of their formulation, with references to more detailed literature on the subject. Although there exist quasi-analytic solutions for some special cases—see [51], [19, Chapter 16] and [82], for examples—numerical analysis remains by far the dominant approach to solving problems of this type. Throughout the chapter, we assume that the reservoir is isothermal, that is, the temperature of the reservoir is uniform in space and constant in time.

Table 7.1 Phases in a compositional model.

Phase	Index α
Aqueous liquid	W
Hydrocarbon liquid	L
Hydrocarbon vapor	V
Solid rock matrix	R

7.1 General Compositional Equations

7.1.1 Constituents, Species, and Phases

Consider a multiphase, multispecies continuum having the four phases listed in Table 7.1. All three fluid phases may be present in a petroleum reservoir, in the form of aqueous formation brine (W), liquid oil (L), and natural-gas vapor (V). Assume that the solid phase R is inert, so that the rock neither dissolves into fluid phases nor adsorbs mass from them.

We allow $N + 1$ chemical species, indexed as $i = 1, 2, \dots, N, \ell$. The species $i = 1, 2, \dots, N$ represent organic species, such as those found in the hydrocarbon fluids L and V, and water. The index ℓ stands for the rock species. Clearly, ℓ must be a **pseudospecies**, lumping together the minerals in the reservoir rock. In practice, the organic species and water, $i = 1, 2, \dots, N$, must also be pseudospecies, since it is impossible to track the flow and transport of the hundreds of thousands of molecular species that comprise any naturally occurring oil or formation brine. Methods for lumping molecular species into pseudospecies include grouping the molecular species by molecular weights or carbon numbers, perhaps assigning species such as CO_2 and CH_4 to their own pseudospecies based on their distinctive dissolution and evaporation properties.

To apply the tools of continuum mechanics to this system, we identify $4(N + 1)$ constituents, each consisting of an ordered pair (i, α), where i represents the species and α represents the phase in which it resides. For example, if carbon dioxide (CO_2) has species index $i = 1$ and methane (CH_4) has index $i = 2$, then the constituent $(1, V)$ denotes carbon dioxide in the hydrocarbon vapor phase, while $(1, L)$ stands for a different constituent, namely carbon dioxide in the hydrocarbon liquid, and $(2, L)$ stands for methane in the hydrocarbon liquid.

Each constituent (i, α) has a true density $\gamma_{(i,\alpha)}$, having dimension [mass of i in α/volume of α], and a velocity $\mathbf{v}_{(i,\alpha)}$. The assumption that the rock is inert implies that $\gamma_{(\ell,\alpha)} = 0$ for $\alpha \neq R$ and that $\gamma_{(i,R)} = 0$ for $i = 1, 2, \dots, N$. Each phase α has a volume fraction ϕ_α, with dimension [volume of α/total volume]. We define the

Table 7.2 Derived quantities used in compositional modeling.

Symbol	Name	Definition
γ_α	Mass density of phase α	$\displaystyle\sum_{i=1}^{N} \gamma_{(i,\alpha)}$
$\omega_{(i,\alpha)}$	Mass fraction of species i in phase α	$\gamma_{(i,\alpha)}/\gamma_\alpha$
ρ	Bulk density of fluids	$\displaystyle\phi\sum_{\alpha\neq R} S_\alpha\gamma_\alpha$
ω_i	Mass fraction of i in fluids	$\displaystyle\frac{\phi}{\rho}\sum_{\alpha\neq R} S_\alpha\gamma_\alpha\omega_{(i,\alpha)}$
$\mathbf{v}_\alpha$	Barycentric velocity of phase α	$\displaystyle\frac{1}{\gamma_\alpha}\sum_{i=1}^{N} \gamma_{(i,\alpha)}\mathbf{v}_{(i,\alpha)}$
$\mathbf{v}_{(i,\alpha)}$	Diffusion velocity of (i,α)	$\mathbf{v}_{(i,\alpha)} - \mathbf{v}_\alpha$

porosity and fluid-phase saturations as usual:

$$\phi = 1 - \phi_R \qquad \text{Porosity,}$$
$$S_\alpha = \phi_\alpha/\phi \qquad \text{Saturation of phase } \alpha, \ \alpha = W, L, V.$$

Based on these quantities, we define a set of derived quantities listed in Table 7.2. These quantities obey the following restrictions:

- Mass fractions sum to 1 in each fluid phase and overall: For $\alpha = W, L, V$,

$$\sum_{i=1}^{N} \omega_{(i,\alpha)} = \sum_{i=1}^{N} \omega_i = 1.$$

- Volume fractions sum to 1:

$$\sum_{\alpha} \phi_\alpha = \sum_{\alpha\neq R} S_\alpha = 1.$$

- Diffusion velocities sum to $\mathbf{0}$ in each fluid phase α:

$$\sum_{i=1}^{N} \mathbf{v}_{(i,\alpha)} = \mathbf{0}.$$

7.1.2 Mass Balance Equations

With the definitions given above, the mass balance for species i in fluid phase α has the following form:

$$\frac{\partial}{\partial t}\left(\phi S_\alpha\gamma_{(i,\alpha)}\right) + \nabla\cdot\left(\phi S_\alpha\gamma_{(i,\alpha)}\mathbf{v}_{(i,\alpha)}\right) = r_{(i,\alpha)}, \tag{7.1}$$

for $i = 1, 2, \ldots, N$ and $\alpha = W, L, V$. (The assumption that the rock species is inert obviates the need to solve a mass balance for species ℓ.) Here, $r_{(i,\alpha)}$ stands for the rate of mass exchange into constituent (i, α) from other constituents. Overall mass balance for the system requires

$$\sum_{i=1}^{N+1} \sum_{\alpha} r_{(i,\alpha)} = 0.$$

Since the solid phase R is inert and consists only of species ℓ, $r_{(\ell,\alpha)} = 0$ for all phases, and $r_{(i,R)} = 0$ for all species. In this case,

$$\sum_{i=1}^{N} \sum_{\alpha \neq R} r_{(i,\alpha)} = 0.$$

In many compositional processes, there are no significant stoichiometric chemical reactions, that is, no reactions that convert mass from a species i to mass of another chemical species. The only reactions that occur are phase exchanges, as when a species in fluid phase V dissolves in fluid phase W or a species in fluid phase L evaporates into fluid phase V. Consequently, there is no net production or loss of any species:

$$\sum_{\alpha \neq R} r_{(i,\alpha)} = 0, \quad i = 1, 2, \ldots, N. \tag{7.2}$$

Exercise 7.1 *Rewrite Eq. (7.1) using phase velocities, then sum over all fluid phases, using Eq. (7.2), to get the following mass balance equation for species $i = 1, 2, \ldots, N$:*

$$\frac{\partial}{\partial t}(\rho \omega_i) + \nabla \cdot \left[\phi \left(S_W \gamma_W \omega_{(i,W)} \mathbf{v}_W + S_L \gamma_L \omega_{(i,L)} \mathbf{v}_L + S_V \gamma_V \omega_{(i,V)} \mathbf{v}_V \right) \right]$$
$$+ \nabla \cdot \left(\mathbf{j}_{(i,W)} + \mathbf{j}_{(i,L)} + \mathbf{j}_{(i,V)} \right) = 0, \quad i = 1, 2, \ldots, N. \tag{7.3}$$

Here,

$$\mathbf{j}_{(i,\alpha)} = \phi S_\alpha \gamma_\alpha \omega_{(i,\alpha)} \mathbf{v}_{(i,\alpha)}$$

*stands for the **diffusive flux** of species i in phase α.*

7.1.3 Species Flow Equations

To obtain flow equations for each species, we adopt the Muskat–Meres extension (6.12) of Darcy's law for the fluid velocities $\mathbf{v}_\alpha$ in Eq. (7.3):

$$\phi S_\alpha \mathbf{v}_\alpha = -\lambda_\alpha (\nabla p_\alpha - \gamma_\alpha g \nabla Z), \quad \alpha = W, L, V.$$

Here,

$$\lambda_\alpha = \frac{k \, k_{r\alpha}}{\mu_\alpha}$$

denotes the mobility of fluid phase α.

Petroleum engineers commonly neglect the diffusive fluxes $\mathbf{j}_{(i,\alpha)}$, reasoning that the transport attributable to applied pressure gradients dominates the transport attributable to hydrodynamic dispersion. This assumption may be reasonable in oilfield operations. The assumption may not be as appropriate in applications to groundwater contamination by NAPL. In this setting, diffusion in the vapor phase may play a significant role in the transport of toxic volatile compounds, small concentrations of which may be of concern. Realistic or not, the assumption leads to the following flow equation for each species i in the fluids:

$$\frac{\partial}{\partial t}\left[\phi\left(S_W\gamma_W\omega_{(i,W)} + S_L\gamma_L\omega_{(i,L)} + S_V\gamma_V\omega_{(i,V)}\right)\right]$$
$$- \nabla \cdot [\lambda_W\gamma_W\omega_{(i,W)}\left(\nabla p_W - \gamma_W g\nabla Z\right) + \lambda_L\gamma_L\omega_{(i,L)}\left(\nabla p_L - \gamma_L g\nabla Z\right)$$
$$+ \lambda_V\gamma_V\omega_{(i,V)}\left(\nabla p_V - \gamma_V g\nabla Z\right)], \quad i = 1, 2, \ldots, N. \tag{7.4}$$

The accumulation and flux terms in each equation allow for the transport of species i in each of the three fluid phases $\alpha = W, L, V$.

Closing this system requires several categories of additional relationships. The first category consists of the restrictions

$$\sum_{i=1}^{N}\omega_{(i,\alpha)} = 1, \text{ for } \alpha = W, L, V; \quad \sum_{i=1}^{N}\omega_i = \sum_{\alpha\neq R}S_\alpha = 1.$$

We often treat the porosity ϕ and depth Z as known functions of position.

Next, for three fluid phases, there are two independent capillary pressure functions, for which the following functional relationships have some empirical justification [164]:

$$p_L = p_W + p_{CLW}(S_W),$$
$$p_V = p_L + p_{CVL}(S_W, S_V).$$

For the relative permeabilities, it is common to adopt the Corey model [40], introduced in Section 6.6, for water-wet rock–fluid systems:

$$k_{rW} = k_{rW}(S_W),$$
$$k_{rL} = k_{rL}(S_W, S_V),$$
$$k_{rV} = k_{rV}(S_V).$$

Finally, we need functional relationships to determine the compositions and saturations of fluid phases W, L, and V, given the overall fluid composition

$$\boldsymbol{\omega} = (\omega_1, \omega_2, \ldots, \omega_N) \tag{7.5}$$

and a fluid pressure, such as p_V. Then, we must calculate the true density γ_α of each fluid phase given its composition

$$\boldsymbol{\omega}_\alpha = (\omega_{(1,\alpha)}, \omega_{(2,\alpha)}, \ldots, \omega_{(N,\alpha)})$$

and pressure p_α. Conceptually, these relationships have the form

$$\omega_\alpha = \omega_\alpha(\omega, p_V),$$
$$S_\alpha = S_\alpha(\omega, p_V),$$
$$\gamma_\alpha = \gamma_\alpha(\omega_\alpha, p_\alpha), \tag{7.6}$$

all for $\alpha = W, L, V$. In practice, these relationships often take the form of systems of nonlinear algebraic equations that define the functions in Eqs. (7.6) implicitly. Section 7.4 provides a brief glimpse into this technically challenging aspect of compositional modeling.

7.2 Black-oil Models

A stalwart among oil reservoir models, the **black-oil** or **beta** model is a special type of compositional model in which the fluid-phase properties depend only on pressure. In addition, the fluid-phase compositions obey a highly restrictive set of interphase mass transfer rules. Although these assumptions seem quite limiting, the black-oil model works quite well for reservoir flows involving nonvolatile oils and no significant changes in interfacial tension among the fluid phases.

7.2.1 Reservoir and Stock-tank Conditions

We allow all three fluid phases W, L, and V to be present. We admit only three pseudospecies. To identify them, petroleum engineers distinguish between **reservoir conditions**—the constant temperature and variable pressure characteristic of the underground formation—and **stock-tank conditions**, defined as 15 °C and 101.325 kPa (1 atm). For most oil reservoirs, reservoir temperatures and pressures are higher than stock-tank conditions. Therefore, the densities of phases W and V at stock-tank conditions are typically lower than at reservoir conditions. In addition, a parcel of the liquid hydrocarbon phase L brought from reservoir conditions to lower-pressure stock-tank conditions will typically release some dissolved natural gas from solution.

To account for the effects of gas dissolving in oil, we identify the following pseudospecies, based on the fluids that are present at stock-tank conditions:

- Species g, a pseudospecies whose composition is the composition of the reservoir gas at stock-tank conditions.
- Species o, a pseudospecies whose composition is the composition of the hydrocarbon liquid at stock-tank conditions.
- Species w, a pseudospecies whose composition is the composition of formation brine at stock-tank conditions.

Table 7.3 Partitioning of pseudospecies in a black-oil model.

Pseudospecies	Mass fraction in W	Mass fraction in L	Mass fraction in V
g	0	$\omega_{(g,L)}$	1
o	0	$\omega_{(o,L)}$	0
w	1	0	0
Sum	1	1	1

We denote the densities of these species at stock-tank conditions by γ_g^{STC}, γ_o^{STC}, and γ_w^{STC}, respectively.

In a black-oil model, these three species partition among the fluid phases W, L, and V at reservoir conditions according to the rules summarized in Table 7.3. The only interphase mass transfers allowed under these rules involve species g dissolving in and evaporating from the liquid hydrocarbon phase L as the pressure changes. To quantify this mass exchange, engineers define the **solution gas–oil ratio**: For a parcel of phase L at reservoir conditions,

$$R_S(p_L) = \frac{\text{STC volume of } g \text{ dissolved in phase } L}{\text{STC volume of } o \text{ in phase } L}.$$

Exercise 7.2 *Show that*

$$\omega_{(g,L)} = \frac{\kappa}{1 + \kappa}, \quad \omega_{(o,L)} = \frac{1}{1 + \kappa},$$

where $\kappa = R_S \gamma_g^{STC} / \gamma_o^{STC}$.

7.2.2 The Black-oil Equations

With these definitions and assumptions, the flow equations for the three species g, o, and w take forms that are simpler than the general flow equations (7.4) developed in Section 7.1. The simplest case is the flow equation for the species w. Since this species resides only in the aqueous phase W, in which $\omega_{(w,W)} = 1$,

$$\frac{\partial}{\partial t}\left(\phi S_W \gamma_W\right) - \nabla \cdot \left[\lambda_W \gamma_W \left(\nabla p_W - \gamma_W g \nabla Z\right)\right] = 0. \tag{7.7}$$

Now relate the true density γ_W of phase W at reservoir conditions to the true density γ_w^{STC} of species w at stock-tank conditions by writing

$$\gamma_W = \frac{\gamma_w^{STC}}{B_W(p_W)}.$$

The factor $B_W(p_W)$ is the **formation volume factor** of the aqueous phase, a dimensionless, empirically measured function that is closely related to the compressibility of the aqueous phase. Substituting for γ_W and dividing through by the constant γ_w^{STC} reduces Eq. (7.7) to the **black-oil equation for water**:

$$\boxed{\frac{\partial}{\partial t}\left(\frac{\phi S_W}{B_W}\right) - \nabla \cdot \left[\frac{\lambda_W}{B_W}\left(\nabla p_W - \gamma_W g \nabla Z\right)\right] = 0.} \qquad (7.8)$$

In this equation, the dimension M does not appear in the accumulation term (the term involving $\partial/\partial t$). Motivated by this observation, engineers often refer to Eq. (7.8) and its analogs for species o and g using the misnomer "volume balance equations."

The flow equation for species o also involves accumulation and flux only in one fluid phase, the hydrocarbon liquid L. However, since the gas species g can be transported in phase L, $\omega_{(o,L)}$ may differ from 1:

$$\frac{\partial}{\partial t}\left(\phi S_L \gamma_L \omega_{(o,L)}\right) - \nabla \cdot \left[\lambda_L \gamma_L \omega_{(o,L)}\left(\nabla p_W + \nabla p_{CLW} - \gamma_L g \nabla Z\right)\right] = 0.$$

Paralleling the derivation for species w, we relate the density of phase L to the densities of its constituents at stock-tank conditions:

$$\gamma_L = \frac{\gamma_o^{STC} + R_S(p_L)\gamma_g^{STC}}{B_L(p_L)}.$$

Here, $B_L(p_L)$ is the empirically measured formation volume factor for the liquid hydrocarbon phase.

Exercise 7.3 *In Eq. (7.4), substitute for γ_L and simplify to get the **black-oil equation for oil**,*

$$\boxed{\frac{\partial}{\partial t}\left(\frac{\phi S_L}{B_L}\right) - \nabla \cdot \left[\frac{\lambda_L}{B_L}\left(\nabla p_W + \nabla p_{CLW} - \gamma_L g \nabla Z\right)\right] = 0.} \qquad (7.9)$$

Finally, the flow equation for the gas species g must track the transport of species g in both the vapor and liquid hydrocarbon phases.

Exercise 7.4 *Begin with the flow equation*

$$\frac{\partial}{\partial t}[\phi(S_L \gamma_L \omega_{(g,L)} + S_V \gamma_V \omega_{(g,V)})]$$
$$-\nabla \cdot [\lambda_L \gamma_L \omega_{(g,L)}(\nabla p_W + \nabla p_{CLW} - \gamma_L g \nabla Z)$$
$$+ \lambda_V \gamma_V \omega_{(g,V)}(\nabla p_W + \nabla p_{CLW} + \nabla p_{CVL} - \gamma_V g \nabla Z)] = 0.$$

Define the formation volume factor for the vapor phase V by the relationship

$$\gamma_V = \frac{\gamma_g^{STC}}{B_V}$$

and use the fact that $\omega_g^V = 1$ *to get the* **black-oil equation for gas,**

$$\frac{\partial}{\partial t}\left[\phi\left(\frac{S_L R_S}{B_L} + \frac{S_V}{B_V}\right)\right]$$

$$-\nabla \cdot \left[\frac{\lambda_L R_S}{B_L}\left(\nabla p_W + \nabla p_{CLW} - \gamma_L g \nabla Z\right)\right. \tag{7.10}$$

$$\left.+ \frac{\lambda_V}{B_V}\left(\nabla p_W + \nabla p_{CLW} + \nabla p_{CVL} - \gamma_V g \nabla Z\right)\right] = 0.$$

Equations (7.8), (7.9), and (7.10) furnish three coupled, nonlinear PDEs governing the transport of the pseudospecies w, o, and g, respectively. The flow equation for g is more complicated than the equations for w and o, because the pseudospecies g can reside, at nonzero mass fractions, in both the vapor phase V and the hydrocarbon liquid phase L. There are many numerical methods for solving this coupled system, including extensions of the simultaneous solution (SS) and the IMPES methods briefly introduced in Section 6.3. See [11,35,115] for details.

7.3 Compositional Flows in Porous Media

Although the black-oil model discussed in Section 7.2 enjoys a reputation as the workhorse of petroleum reservoir simulation, in many oil recovery processes, such as miscible gas flooding, more significant effects of composition on fluid-phase properties play critical roles in the physics [131]. This section examines a set of flow equations that model these phenomena. Section 7.4 sketches the fluid-phase thermodynamics that determine fluid-phase compositions.

7.3.1 A Simplified Compositional Formulation

For simplicity's sake, we restrict attention to compositional flows in which interphase mass transfer is limited to mass exchanges between the hydrocarbon liquid phase L and the hydrocarbon vapor phase V. This assumption is unrealistic, because it neglects the dissolution of species such as methane and carbon dioxide in the aqueous phase W. However, it allows us to examine the thermodynamic principles in a simple setting that involves two-phase vapor–liquid equilibrium.

If we label the fluid-phase species $i = 1, 2, \ldots, N$ so that species N is the water pseudospecies w, as in Section 7.2, the assumption above implies that

$$\omega_{(i,W)} = 0, \qquad i = 1, 2, \ldots, N - 1;$$

$$\omega_{(N,L)} = \omega_{(N,V)} = 0.$$

Therefore, $\omega_{(N,W)} = 1$, and the flow equation for water reduces to the following PDE:

$$\frac{\partial}{\partial t} \left(\phi S_W \gamma_W \right) - \nabla \cdot \left[\lambda_W \gamma_W \left(\nabla p_L - \nabla p_{CLW} - \gamma_W g \nabla Z \right) \right] = 0.$$

Remaining are the $N - 1$ flow equations for the species residing in the hydrocarbon liquid and vapor phases:

$$\frac{\partial}{\partial t} \left(\phi \gamma_H \omega_{(i,H)} \right) - \nabla \cdot \left[\lambda_L \gamma_L \omega_{(i,L)} \left(\nabla p_L - \gamma_L g \nabla Z \right) \right]$$
$$- \nabla \cdot \left[\lambda_V \gamma_V \omega_{(i,V)} \left(\nabla p_L + \nabla p_{CVL} - \gamma_V g \nabla Z \right) \right] = 0, \qquad (7.11)$$

for $i = 1, 2, \ldots, N - 1$. Here,

$$\gamma_H = S_L \gamma_L + S_V \gamma_V$$

stands for the total density of the hydrocarbon fluids, and

$$\omega_{(i,H)} = \frac{S_L \gamma_L \omega_{(i,L)} + S_V \gamma_V \omega_{(i,V)}}{\gamma_H}$$

denotes the total mass fraction of species i in the hydrocarbon fluids.

7.3.2 Conversion to Molar Variables

Petroleum engineers commonly write compositional models in terms of molar quantities. This reformulation facilitates the thermodynamic calculations required in fluid-phase behavior calculations outlined in Section 7.4. For this purpose, given $N - 1$ species or pseudospecies in the hydrocarbon fluid phases L and V, Table 7.4 defines molar variables corresponding to the mass-related variables.

These quantities obey the following restrictions:

$$\sum_{i=1}^{N-1} \tilde{\omega}_{(i,\alpha)} = \sum_{i=1}^{N-1} \tilde{\omega}_{(i,H)} = Y_L + Y_V = 1.$$

Once we know the molar densities $\tilde{\gamma}_L$ and $\tilde{\gamma}_V$ of the hydrocarbon phases, the phase mole fractions Y_L and Y_V stand in direct correspondence to the saturations S_L and S_V, as the following exercise shows.

Exercise 7.5 *Show that*

$$S_L = \frac{Y_L \tilde{\gamma}_V}{Y_L \tilde{\gamma}_V + Y_V \tilde{\gamma}_L}, \qquad S_V = \frac{Y_V \tilde{\gamma}_L}{Y_L \tilde{\gamma}_V + Y_V \tilde{\gamma}_L}. \qquad (7.12)$$

Table 7.4 Molar quantities used in compositional reservoir modeling.

Symbol	Name	Definition
$\tilde{M}_i$	Molar mass of species i	Mass of i per mole of i
$\tilde{\gamma}_{(i,\alpha)}$	Molar density of species i	$\gamma_{(i,\alpha)}/\tilde{M}_i$
$\tilde{\gamma}_\alpha$	Molar density of phase α	$\displaystyle\sum_{i=1}^{N-1} \tilde{\gamma}_{(i,\alpha)}$
$\tilde{\omega}_{(i,\alpha)}$	Mole fraction of species i in phase α	$\tilde{\gamma}_{(i,\alpha)}/\tilde{\gamma}_\alpha$
$\tilde{\rho}_H$	Bulk molar density of hydrocarbons	$\phi(S_L\tilde{\gamma}_L + S_V\tilde{\gamma}_V)$
$\tilde{\omega}_{(i,H)}$	Mole fraction of species i in hydrocarbons	$\dfrac{S_L\tilde{\gamma}_L\tilde{\omega}_{(i,L)} + S_V\tilde{\gamma}_V\tilde{\omega}_{(i,V)}}{\tilde{\rho}_H}$
Y_α	Mole fraction of phase α in hydrocarbons	$\dfrac{S_\alpha\tilde{\gamma}_\alpha}{S_L\tilde{\gamma}_L + S_V\tilde{\gamma}_V}$

Multiplying each hydrocarbon species flow equation (7.11) by $1/\tilde{M}_i$ and using the definitions in Table 7.4 yield the following molar form of the species flow equations:

$$\frac{\partial}{\partial t}\left(\tilde{\gamma}_{(i,H)}\tilde{\omega}_{(i,H)}\right) - \nabla \cdot \left[\lambda_L\tilde{\gamma}_L\tilde{\omega}_{(i,L)}(\nabla p_L + -\gamma_L\nabla Z)\right]$$
$$- \nabla \cdot \left[\lambda_V\tilde{\gamma}_V\tilde{\omega}_{(i,V)}(\nabla p_L + \nabla p_{CVL} - \gamma_V\nabla Z)\right] = 0, \qquad (7.13)$$

for $i = 1, 2, \ldots, N - 1$.

Summing over the species $i = 1, 2, \ldots, N - 1$ gives a molar form of the total hydrocarbon flow equation,

$$\frac{\partial\tilde{\rho}_H}{\partial t} - \nabla \cdot [(\lambda_L\tilde{\gamma}_L + \lambda_V\tilde{\gamma}_V)\nabla p_L + \lambda_V\tilde{\gamma}_V\nabla p_{CVL}$$
$$- (\lambda_L\tilde{\gamma}_L\gamma_L + \lambda_V\tilde{\gamma}_V\gamma_V)\nabla Z] = 0. \qquad (7.14)$$

Some numerical formulations utilize Eq. (7.14) as a pressure equation, leaving $N - 2$ independent equations of the form (7.13) to update the species mole fractions $\tilde{\omega}_{(i,H)}$ at each time step. This approach parallels that used in IMPES methods for two-phase flow and black-oil models, as introduced in Section 6.3. For examples of different numerical formulations of compositional models, including analogs of the simultaneous solution (SS) method, see [2,38,56,109,149,162].

7.4 Fluid-phase Thermodynamics

As discussed in Section 7.1, closing the system (7.13) of flow equations requires additional relationships. In particular, we must determine

- the molar compositions of each hydrocarbon phase:

$$\tilde{\omega}_\alpha = \left(\tilde{\omega}_{(1,\alpha)}, \tilde{\omega}_{(2,\alpha)}, \ldots, \tilde{\omega}_{(N-1,\alpha)} \right), \quad \text{for } \alpha = L, V,$$

- the saturations S_L and S_V,
- the molar densities $\tilde{\gamma}_L, \tilde{\gamma}_V$,

given the overall hydrocarbon mole fractions

$$\tilde{\omega} = \left(\tilde{\omega}_{(1,H)}, \tilde{\omega}_{(2,H)}, \ldots, \tilde{\omega}_{(N-1,H)} \right).$$

In practice, these additional relationships do not take the explicit forms that Eqs. (7.6) suggest. Instead, they commonly take the form of thermodynamic constraints, which furnish a system of nonlinear algebraic equations that define the functional relationships (7.6) implicitly. This system must be solved numerically, using an iterative method such as Newton's method [5, Chapter 4]. This section sketches the origins of this system of equations and gives a brief description of their consequences in certain types of compositional flows.

7.4.1 Flash Calculations

The thermodynamic constraints arise from the assumption that the fluid phases remain in **local thermodynamic equilibrium**. According to this assumption, the time needed for the two phases L and V to reach their equilibrium compositions, saturations, and densities after mixing is much shorter than the time scales associated with the species transport. As a consequence, at each point in space and time, the two fluids are in thermodynamic equilibrium with each other. This assumption enables us to use principles of equilibrium thermodynamics elucidated in 1878 by American mathematician J. Willard Gibbs [60] and advanced significantly during the ensuing century; see Prausnitz et al. [125] for an encyclopedic reference.

At the core of vapor–liquid equilibrium calculations is a set of constraints requiring that the **fugacities** $f_{(i,L)}$ and $f_{(i,V)}$ of each species i in the fluid phases L and V be equal. The fugacity generalizes the concept of the partial pressure in an ideal gas. The equal-fugacity constraints $f_{(i,L)} = f_{(i,V)}$ arise from a minimization principle derived by Gibbs. Section 7.4.2 reviews one approach for calculating fugacities in isothermal hydrocarbon mixtures given phase compositions and pressures.

Figure 7.1 depicts the overall logic of the thermodynamic calculations. There are two steps. The first is more computationally intensive: Given the pressures p_L

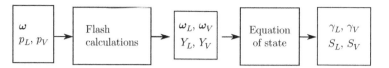

Figure 7.1 Flow chart for equation-of-state thermodynamic calculations in a compositional model.

and p_V and the array $\tilde{\omega}$ of overall mole fractions, determine the compositions and mole fractions of the two hydrocarbon fluid phases,

$$\tilde{\omega}_L, \tilde{\omega}_L, Y_L, Y_V. \tag{7.15}$$

This step uses a procedure commonly referred to as the **flash calculation**. It determines the $2N$ unknowns listed in (7.15) by solving a nonlinear system of $2N$ algebraic equations that includes the following:

1. $N-1$ equal-fugacity constraints, having the form

$$f_{(i,L)}(\tilde{\omega}_L, p_L) - f_{(i,V)}(\tilde{\omega}_V, p_V) = 0, \quad i = 1, 2, \ldots, N-1. \tag{7.16}$$

2. $N-2$ restrictions on the species mole fractions:

$$Y_L \tilde{\omega}_{(i,L)} + Y_V \tilde{\omega}_{(i,V)} = \tilde{\omega}_{(i,H)}, \quad i = 1, 2, \ldots, N-2. \tag{7.17}$$

3. A restriction derived by summing the restrictions (7.17) for $i = 1, 2, \ldots, N-1$:

$$Y_L + Y_V = 1.$$

4. Two restrictions on phase compositions:

$$\sum_{i=1}^{N-1} \tilde{\omega}_{(i,L)} = 1 = \sum_{i=1}^{N-1} \tilde{\omega}_{(i,V)}.$$

The second step in the thermodynamic calculations is to determine the true densities γ_L and γ_V and saturations S_L and S_V of the two hydrocarbon phases. In an isothermal continuum, given the phase compositions and pressure of each phase, one can determine the phase molar densities using an equation of state, as sketched in Section 7.4.2 Once we know the phase molar densities of phases L and V, we compute their saturations using Eqs. (7.12).

7.4.2 Equation-of-state Methods

A common methodology for implementing these thermodynamic calculations is to use **equation-of-state** methods. Equations of state relate a fluid's pressure, molar density, composition, and temperature. The development of such equations follows a venerable tradition beginning with Dutch physicist Johannes van der Waals in 1873; it requires intimate knowledge of molecular thermodynamics as well as an instinct for curve-fitting to model the behavior of gases and liquids. This section provides only a brief overview of this highly technical field. For detailed examples of equation-of-state methods, see [38,57,100,101,109].

In the context of hydrocarbon fluid mixtures, it is common to write the equation of state for a fluid phase α in terms of its dimensionless **compressibility factor**,

$$Z_\alpha = \frac{p_\alpha}{\tilde{\gamma}_\alpha R_{gas} T}.$$

Here, T denotes the temperature, and R_{gas} is the gas constant, $8.31434\,\mathrm{J\,mol^{-1}\,K^{-1}}$ in SI units. Since $Z_\alpha = 1$ when the phase α is an ideal gas, the compressibility factor effectively measures the departure of the fluid's density–pressure relationship from ideality.

Among the most commonly used equations of state are the cubic equations of Redlich and Kwong [126], Soave [137], and Peng and Robinson [116]. These equations have the form

$$Z_\alpha^3 + c_2 Z_\alpha^2 + c_1 Z_\alpha + c_0 = 0,$$

where the coefficients c_0, c_1, and c_2 are parameters determined by the composition, pressure, and temperature of phase α. By calculating these coefficients and finding the corresponding roots Z_α for the fluid phases L and V, one can compute the molar densities $\tilde{\gamma}_L$ and $\tilde{\gamma}_V$. Then, using the phase mole fractions Y_L and Y_V, one can determine the saturations S_L and S_V using the results of Exercise 7.5.

One attractive feature of equation-of-state methods is their internal consistency. Each equation of state furnishes an expression for the fugacity $f_{(i,\alpha)}$ of species i in phase α; see Peng and Robinson [116] for an example. This fact allows us to develop flash calculations that maintain mathematical consistency with the calculations of phase molar densities and saturations.

Fluid-phase thermodynamics impose heavy computational burdens on compositional simulators. Flash calculations are especially time consuming, since the equal-fugacity constraints (7.16) require iterative methods for solving nonlinear algebraic systems. A compositional simulator must execute these calculations in every spatial grid cell at every time step and, typically, at every iteration of the iterative time-stepping method required to solve the nonlinear species flow equations (7.13).

Appendix A

Dedicated Symbols

This appendix lists symbols used to denote specific physical quantities. In many cases, the main text employs subscripts, superscripts, and accent marks for particular instances of these quantities; the table below makes no attempt to list all of these particular cases.

Throughout the book, vectors appear in boldface font; tensors appear in bold sans-serif font.

Table A.1   Dedicated symbols for physical quantities.

Symbol	Physical quantity	Dimension
B	Formation volume factor	1
c	Concentration	ML^{-3}
		1 if normalized
D	Diffusion coefficient	L^2T^{-1}
D	Hydrodynamic dispersion tensor	L^2T^{-1}
f	Fractional flow	1
$F,\mathbf{F}$	Flux or force	LT^{-1} or MLT^{-2}
g	Gravitational acceleration	LT^{-2}
H	Piezometric head	L
I	Identity tensor	1
$\mathbf{j}$	Diffusive flux	$ML^{-2}T^{-1}$
k,k	Permeability	L^2
K,K	Hydraulic conductivity	LT^{-1}
$\log$	Natural logarithm	1

The Mathematics of Fluid Flow Through Porous Media, First Edition. Myron B. Allen.
© 2021 John Wiley & Sons, Inc. Published 2021 by John Wiley & Sons, Inc.

Table A.1 (Continued)

Symbol	Physical quantity	Dimension
p	Pressure	$ML^{-1}T^{-2}$
Pe	Péclet number	1
Q	Volumetric flow rate	L^3T^{-1}
r	Radial coordinate	L
R_S	Solution gas–oil ratio	1
Re	Reynolds number	1
$\bar{s}$	Drawdown	L
S	Fluid saturation or storativity	1
S_p	Storage	$M^{-1}LT^2$
S_s	Specific storage	L^{-1}
t	Time	T
T	Transmissivity	L^2T^{-1}
T	Stress	$ML^{-1}T^{-2}$
$\mathbf{v}$	Average interstitial velocity	LT^{-1}
$x, \mathbf{x}$	Spatial position	L
z	Axial coordinate	L
Z	Depth	L
Z_α	Compressibility factor of phase α	1
γ	True density	ML^{-3}
δ	Dirac δ distribution	1
θ	Angle or azimuth	Radians
Θ	Moisture content	1
λ	Fluid mobility	$M^{-1}L^3T$
μ	Dynamic viscosity	$ML^{-1}T^{-1}$
ν	Kinematic viscosity	L^2T^{-1}
ρ	Density	ML^{-3}
σ	Surface tension	MT^{-2}
ϕ	Porosity	1
$\phi\mathbf{v}$	Filtration velocity or specific discharge	LT^{-1}
$\hat{\phi}$	Polar angle	Radians
Ψ	Pressure head or tension head	L
ω	Mass fraction	1
$\| \mathbf{x} \|$	Euclidean length, $\sqrt{\mathbf{x} \cdot \mathbf{x}}$	L

Appendix B

Useful Curvilinear Coordinates

Although many treatments gloss over the representations of differential operators in polar, cylindrical, and spherical coordinates, the conversions are worth reviewing, since Chapters 2 and 4 use the results.

B.1 Polar Coordinates

In two space dimensions, the following transformation defines polar coordinates in terms of the standard Cartesian coordinates (x_1, x_2):

$$\Phi\left(\begin{bmatrix} r \\ \theta \end{bmatrix}\right) = \begin{bmatrix} r\cos\theta \\ r\sin\theta \end{bmatrix} = \begin{bmatrix} x_1 \\ x_2 \end{bmatrix}. \tag{B.1}$$

The **radial coordinate** $r = \| \mathbf{x} \| = \sqrt{\mathbf{x} \cdot \mathbf{x}}$ represents distance from the origin; θ represents the counterclockwise angle from the x_1-axis.

Exercise B.1

(A) *Show that the transformation* (B.1) *has derivative*

$$\Phi' = \begin{bmatrix} \partial x_1/\partial r & \partial x_1/\partial\theta \\ \partial x_2/\partial r & \partial x_2/\partial\theta \end{bmatrix} = \begin{bmatrix} \cos\theta & -r\sin\theta \\ \sin\theta & r\cos\theta \end{bmatrix},$$

   with **Jacobian determinant**

$$\det \Phi' = r.$$

(B) *Apply the chain rule to the composition $\varphi(\Phi(r,\theta))$ to show that the two-dimensional gradient $\nabla\varphi$ of a differentiable function has matrix representation*

$$\left[\frac{\partial\varphi}{\partial x_1}, \frac{\partial\varphi}{\partial x_2}\right] = \left[\cos\theta\frac{\partial\varphi}{\partial r} - \frac{\sin\theta}{r}\frac{\partial\varphi}{\partial\theta}, \quad \sin\theta\frac{\partial\varphi}{\partial r} + \frac{\cos\theta}{r}\frac{\partial\varphi}{\partial\theta}\right]. \tag{B.2}$$

The Mathematics of Fluid Flow Through Porous Media, First Edition. Myron B. Allen.
© 2021 John Wiley & Sons, Inc. Published 2021 by John Wiley & Sons, Inc.

(C) *Show that the two-dimensional Laplace operator in polar coordinates is*

$$\nabla^2 = \nabla \cdot \nabla = \frac{1}{r}\frac{\partial}{\partial r}\left(r\frac{\partial}{\partial r}\right) + \frac{1}{r^2}\frac{\partial^2}{\partial \theta^2}. \tag{B.3}$$

Transforming integrals from Cartesian coordinates to polar coordinates requires the change-of-variables theorem [99, Section 6.2]. For a transformation $\mathbf{\Phi}$ that maps a region $\mathcal{R}$ of (r, θ)-space into a region $\mathbf{\Phi}(\mathcal{R})$ in the rectangular coordinates (x_1, x_2), this theorem gives

$$\int_{\mathbf{\Phi}(\mathcal{R})} f(x_1, x_2)\, dx_1\, dx_2 = \int_{\mathcal{R}} f(\mathbf{\Phi}(r, \theta))\, |\det \mathbf{\Phi}'(r, \theta)|\, dr\, d\theta$$

$$= \int_{\mathcal{R}} f(x_1(r, \theta), x_2(r, \theta))\, r\, dr\, d\theta, \tag{B.4}$$

for a sufficiently well behaved function f.

B.2 Cylindrical Coordinates

The following coordinate transformation defines **cylindrical coordinates**:

$$\mathbf{\Phi}\left(\begin{bmatrix} z \\ r \\ \theta \end{bmatrix}\right) = \begin{bmatrix} z \\ r\cos\theta \\ r\sin\theta \end{bmatrix} = \begin{bmatrix} x_1 \\ x_2 \\ x_3 \end{bmatrix}. \tag{B.5}$$

Here, x_1, x_2, and x_3 are Cartesian coordinates, and

$$-\infty < z < \infty, \quad 0 \le r < \infty, \quad 0 \le \theta < 2\pi.$$

As Figure B.1 illustrates, the **axial coordinate** z gives the position along an axis, which we take to be the x_1 axis; the **radial coordinate** r gives distance from the axis, and the **azimuth** θ gives the angle with respect to a reference plane passing through the axis.

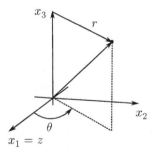

Figure B.1 Cylindrical coordinates.

In the standard orthonormal basis $\{\mathbf{e}_1, \mathbf{e}_2, \mathbf{e}_3\}$ associated with Cartesian coordinates, the derivative of the transformation (B.5) is

$$\mathbf{\Phi}' = \begin{bmatrix} \partial x_1/\partial z & \partial x_1/\partial r & \partial x_1/\partial \theta \\ \partial x_2/\partial z & \partial x_2/\partial r & \partial x_2/\partial \theta \\ \partial x_3/\partial z & \partial x_3/\partial r & \partial x_3/\partial \theta \end{bmatrix} = \begin{bmatrix} 1 & 0 & 0 \\ 0 & \cos\theta & -r\sin\theta \\ 0 & \sin\theta & r\cos\theta \end{bmatrix},$$

with Jacobian determinant

$$\det \mathbf{\Phi}' = r.$$

By the chain rule, for any differentiable function $\varphi(\mathbf{x}(z, r, \theta))$,

$$\left[\frac{\partial\varphi}{\partial z}, \frac{\partial\varphi}{\partial r}, \frac{\partial\varphi}{\partial \theta} \right] = \left[\frac{\partial\varphi}{\partial x_1}, \frac{\partial\varphi}{\partial x_2}, \frac{\partial\varphi}{\partial x_3} \right] \mathbf{\Phi}',$$

which we solve to find the matrix representation of $\nabla\varphi(\mathbf{x}(z, r, \theta))$ in cylindrical coordinates:

$$\left[\frac{\partial\varphi}{\partial x_1}, \frac{\partial\varphi}{\partial x_2}, \frac{\partial\varphi}{\partial x_3} \right] = \left[\frac{\partial\varphi}{\partial z}, \cos\theta\frac{\partial\varphi}{\partial r} - \frac{\sin\theta}{r}\frac{\partial\varphi}{\partial \theta}, \sin\theta\frac{\partial\varphi}{\partial r} + \frac{\cos\theta}{r}\frac{\partial\varphi}{\partial \theta} \right]. \quad (B.6)$$

The left side of Eq. (B.6) gives the coordinates of $\nabla\varphi$ with respect to the basis $\{\mathbf{e}_1, \mathbf{e}_2, \mathbf{e}_3\}$:

$$\nabla\varphi = \sum_{j=1}^{3} \frac{\partial\varphi}{\partial x_j}\mathbf{e}_j.$$

Often of greater utility in the cylindrical coordinate system is the associated basis $\{\mathbf{e}_z, \mathbf{e}_r, \mathbf{e}_\theta\}$ of unit-length, mutually orthogonal vectors:

$$\mathbf{e}_z = \mathbf{e}_3,$$
$$\mathbf{e}_r = (\cos\theta)\mathbf{e}_1 + (\sin\theta)\mathbf{e}_2,$$
$$\mathbf{e}_\theta = (-\sin\theta)\mathbf{e}_1 + (\cos\theta)\mathbf{e}_2.$$

Exercise B.2   *Sketch these vectors, and show that*

$$\nabla\varphi = \frac{\partial\varphi}{\partial z}\mathbf{e}_z + \frac{\partial\varphi}{\partial r}\mathbf{e}_r + \frac{1}{r}\frac{\partial\varphi}{\partial \theta}\mathbf{e}_\theta.$$

Exercise B.3   *Show that, in cylindrical coordinates, the Laplace operator has the following form:*

$$\boxed{\nabla^2 = \nabla\cdot\nabla = \frac{\partial^2}{\partial z^2} + \frac{1}{r}\frac{\partial}{\partial r}\left(r\frac{\partial}{\partial r}\right) + \frac{1}{r^2}\frac{\partial^2}{\partial \theta^2}.} \quad (B.7)$$

To transform integrals from Cartesian coordinates to cylindrical coordinates, we again use the change-of-variables theorem. For a transformation $\mathbf{\Phi}$ that maps a region $\mathcal{R}$ of a generic three-dimensional space, having coordinates (u_1, u_2, u_3), into a region $\Phi(\mathcal{R})$ in the rectangular coordinates (x_1, x_2, x_3), this theorem gives

$$\int_{\Phi(\mathcal{R})} f(x_1, x_2, x_3) \, dx_1 \, dx_2 \, dx_3$$

$$= \int_{\mathcal{R}} f(\Phi(u_1, u_2, u_3)) |\det \Phi'(u_1, u_2, u_3)| \, du_1 \, du_2 \, du_3$$

$$= \int_{\mathcal{R}} f(r\cos\theta, r\sin\theta, z) \, r \, dr \, d\theta \, dz,$$

for any sufficiently well-behaved function f. For example, the integral of a sufficiently well-behaved function f over all of three-dimensional space is

$$\int_{-\infty}^{\infty} \int_{-\infty}^{\infty} \int_{-\infty}^{\infty} f(x_1, x_2, x_3) \, dx_1 \, dx_2 \, dx_3$$

$$= \int_{-\infty}^{\infty} \int_{0}^{2\pi} \int_{0}^{\infty} f(r\cos\theta, r\sin\theta, z) \, r \, dr \, d\theta \, dz.$$

B.3   Spherical Coordinates

The coordinate transformation

$$\Phi\left(\begin{bmatrix} r \\ \theta \\ \hat{\phi} \end{bmatrix}\right) = \begin{bmatrix} r\sin\hat{\phi}\cos\theta \\ r\sin\hat{\phi}\sin\theta \\ r\cos\hat{\phi} \end{bmatrix} = \begin{bmatrix} x_1 \\ x_2 \\ x_3 \end{bmatrix}$$

defines **spherical coordinates**. Here,

$$0 \le r < \infty, \quad 0 \le \theta < 2\pi, \quad 0 \le \phi \le \pi.$$

As shown in Figure B.2, the **radial coordinate** r gives distance from the origin; the **polar angle** $\hat{\phi}$ gives the angle with respect to an axis passing through the origin, and the **azimuth** θ gives the angle with respect to a plane passing through that axis.

Exercise B.4   *Calculate the derivative Φ' in Cartesian coordinates.*

We associate with spherical coordinates the basis $\{\mathbf{e}_r, \mathbf{e}_\theta, \mathbf{e}_\phi\}$, where

$$\mathbf{e}_r = \sin\phi[(\cos\theta)\mathbf{e}_1 + (\sin\theta)\mathbf{e}_2] + (\cos\hat{\phi})\mathbf{e}_3,$$

$$\mathbf{e}_\theta = -(\sin\theta)\mathbf{e}_1 + (\cos\theta)\mathbf{e}_2,$$

$$\mathbf{e}_\phi = \cos\hat{\phi}[(\cos\theta)\mathbf{e}_1 + (\sin\theta)\mathbf{e}_2] - (\sin\hat{\phi})\mathbf{e}_3.$$

Figure B.2   Spherical coordinates.

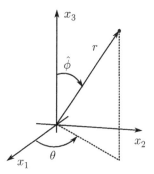

Exercise B.5   *For a differentiable function $\varphi(\mathbf{x}(r, \theta, \phi))$, show that*

$$\nabla\varphi = \frac{\partial\varphi}{\partial r}\mathbf{e}_r + \frac{1}{r\sin\hat{\phi}}\frac{\partial\varphi}{\partial\theta}\mathbf{e}_\theta + \frac{1}{r}\frac{\partial\varphi}{\partial\hat{\phi}}\mathbf{e}_\phi$$

and

$$\nabla^2\varphi = \frac{1}{r^2}\frac{\partial}{\partial r}\left(r^2\frac{\partial\varphi}{\partial r}\right) + \frac{1}{r^2\sin^2\hat{\phi}}\frac{\partial^2\varphi}{\partial\theta^2} + \frac{1}{r^2\sin\hat{\phi}}\frac{\partial}{\partial\hat{\phi}}\left(\sin\hat{\phi}\frac{\partial\varphi}{\partial\hat{\phi}}\right).$$

Exercise B.6   *Use the change-of-variables theorem to establish the following formula for converting an integral from Cartesian coordinates to spherical coordinates:*

$$\int_{\Phi(R)} f(x_1, x_2, x_3)\, dx_1\, dx_2\, dx_3$$

$$= \int_R f(r\sin\hat{\phi}\cos\theta, r\sin\hat{\phi}\sin\theta, r\cos\hat{\phi})\, r^2\sin\hat{\phi}\, dr\, d\theta\, d\hat{\phi},$$

for any sufficiently well behaved function f.

Appendix C

The Buckingham Pi Theorem

This appendix reviews a short proof of a central theorem in dimensional analysis. The proof is often attributed to Edgar Buckingham [31], an American physicist who also played a key role in the early analysis of multifluid flows in porous media, as reviewed in Section 6.2. However, a French mathematician, Joseph Bertrand [22], established the fundamental concept of the theorem in 1878.

C.1 Physical Dimensions and Units

All physical dimensions involve products of powers of basic dimensions, denoted abstractly as $L_1, \dots, L_M$. For us, these basic dimensions are length L, mass M, and time T. A variable is **dimensionless** if its dimension is 1.

Assigning a numerical value to a dimensional variable requires choosing a system of **units** in which to measure variables having dimensions $L_1, \dots, L_M$. For example, the SI (*Système Internationale*) uses meters (m), kilograms (kg), and seconds (s) to measure length, mass, and time, respectively; see Section 1.3. A system of units is **consistent** if, for each index $j = 1, \dots, M$, it measures all occurrences of dimension L_j using the same unit. For example, in SI the unit of velocity is m s^{-1}, and the unit of force is kg m s^{-2} = N, the newton. Both measure length in meters. The numerical value of a dimensionless variable is independent of the choice of units, so long as we use consistent sets of units.

The Buckingham Pi theorem rests on the underlying requirement that any physical law must be **unit-free**. More specifically, consider a physical law of the form

$$\varphi(q_1, \dots, q_m) = 0, \tag{C.1}$$

giving a relationship between dimensional quantities $q_1, \dots, q_m$. To be unit-free, this relationship must hold no matter what consistent set of units we use to measure $q_1, \dots, q_m$. This condition essentially guarantees that physical laws not depend on subjective choices of the units used to measure physical variables.

The Mathematics of Fluid Flow Through Porous Media, First Edition. Myron B. Allen.
© 2021 John Wiley & Sons, Inc. Published 2021 by John Wiley & Sons, Inc.

C.2 The Buckingham Theorem

Theorem C.1 *Buckingham Pi Theorem* *For any unit-free physical law of the form* (C.1) *giving a relationship among dimensional variables $q_1, \ldots, q_m$, there is a set $\Pi_1, \ldots, \Pi_{m-l}$ of dimensionless variables, with $0 \le l < m$, for which an equivalent law of the form*

$$\Phi(\Pi_1, \ldots, \Pi_{m-l}) = 0$$

holds.

Proof: The argument proceeds in two steps. Step 1 is to find the dimensionless variables $\Pi_1, \ldots, \Pi_{m-l}$ and the parameter l. Each dimensionless variable, denoted generically by Π, has the form

$$\Pi = \prod_{j=1}^{m} q_j^{n_j}, \tag{C.2}$$

for some exponents $n_1, \ldots, n_m$. Let $L_1, \ldots, L_k$ be the basic dimensions required to form the variables $q_1, \ldots, q_m$. Thus, $k \le m$, and each dimensional variable q_j has dimension

$$\dim(q_j) = L_1^{a_{1j}} \cdots L_k^{a_{kj}},$$

for some integer powers $a_{1j}, \ldots, a_{kj}$. Therefore,

$$\dim(\Pi) = \prod_{j=1}^{m} \dim(q_j)^{n_j} = \prod_{j=1}^{m} \left(L_1^{a_{1j}} \cdots L_k^{a_{kj}} \right)^{n_j} = \prod_{l=1}^{k} L_l^{n_1 a_{l1} + \cdots + n_m a_{lm}}.$$

But the condition $\dim(\Pi) = 1$ implies that the exponents of $L_1, \ldots, L_k$ all vanish. This condition yields the homogeneous linear system

$$\begin{bmatrix} a_{11} & \cdots & a_{1m} \\ \vdots & & \vdots \\ a_{k1} & \cdots & a_{km} \end{bmatrix} \begin{bmatrix} n_1 \\ \vdots \\ n_m \end{bmatrix} = \begin{bmatrix} 0 \\ \vdots \\ 0 \end{bmatrix}. \tag{C.3}$$

Since $k \le m$, this system is possibly underdetermined. The number of linearly independent (hence nonzero) solution vectors $(n_1, \ldots, n_m)$ is $m - l$, where $l \le m$ denotes the rank of the $k \times m$ matrix in Eq. (C.3). Each of these solutions defines an independent dimensionless variable of the form (C.2). We denote these dimensionless variables as Π_j, for $j = 1, \ldots, m - l$.

Step 2 of the proof is to find a dimensionless law equivalent to Eq. (C.1). Begin by substituting dimensionless variables $\Pi_1, \ldots, \Pi_{m-l}$ for $m - l$ of the dimensional variables $q_1, \ldots, q_m$. Renumbering the dimensional variables if necessary, we deduce a functional relationship that is equivalent to Eq. (C.1) having the form

$$\varphi(q_1, \ldots, q_m) = \Phi(\Pi_1, \ldots, \Pi_{m-l}, q_{m-l+1}, \ldots, q_m) = 0. \tag{C.4}$$

The hypothesis that the physical law (C.1) is unit-free allows us to measure the dimensional quantities $q_{m-l+1}, \ldots, q_m$ in any consistent set of units, getting any numerical values we wish for these arguments of Φ, while the values of the dimensionless quantities $\Pi_1, \ldots, \Pi_{m-l}$ remain unchanged. It follows that Φ cannot depend on the numerical values of $q_{m-l+1}, \ldots, q_m$. Therefore, Eq. (C.4) reduces to the simpler form:

$$\Phi(\Pi_1, \ldots, \Pi_{m-l}) = 0. \qquad\qquad\qquad \blacksquare$$

Appendix D

Surface Integrals

This appendix contains a brief review of surface integration. It covers three topics: (i) the definition of a surface integral; (ii) the statement of the Stokes theorem, which relates a surface integral to an integral over the boundary of the surface; and (iii) a corollary to the Stokes theorem used in deriving the Young–Laplace equation (6.2). For more detailed coverage of these topics, consult [99, Chapters 7-8].

D.1 Definition of a Surface Integral

The definition of a surface integral rests on the concept of parametrizations. A parametrization of a of surface Σ in three-dimensional Euclidean space represents the surface as a smooth image of a region, called the **parameter domain**, in the plane $\mathbb{R}^2$. More precisely, let $\Omega \subset \mathbb{R}^2$ be the parameter domain together with its boundary $\partial\Omega$, as drawn in Figure D.1. A continuously differentiable, one-to-one function φ defined on Ω **parametrizes** the surface Σ if:

1. $\varphi(\Omega) = \Sigma$.
2. The derivative $\varphi'(\xi)$, a 3×2 matrix, has linearly independent columns at every point $\xi = (\xi_1, \xi_2)$ in the parameter domain Ω.

Geometrically, the columns of $\varphi'(\xi)$, given by

$$\tau_1 = \varphi'(\xi) \begin{bmatrix} 1 \\ 0 \end{bmatrix}, \quad \tau_2 = \varphi'(\xi) \begin{bmatrix} 0 \\ 1 \end{bmatrix},$$

are vectors tangent to the surface Σ at each point $\varphi(\xi)$. Condition 2 on φ' ensures that these vectors are noncollinear and hence that their cross product $\tau_1(\xi) \times \tau_2(\xi) \neq \mathbf{0}$. As a result, there exists a continuous, unit-length vector field

$$\mathbf{n}(\xi) = \frac{\tau_1(\xi) \times \tau_2(\xi)}{\| \tau_1(\xi) \times \tau_2(\xi) \|}$$

The Mathematics of Fluid Flow Through Porous Media, First Edition. Myron B. Allen.
© 2021 John Wiley & Sons, Inc. Published 2021 by John Wiley & Sons, Inc.

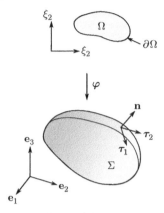

Figure D.1 A parametrized surface $\Sigma = \varphi(\Omega)$ showing the unit normal vector **n** and the tangent vectors τ_1 and τ_2.

that is perpendicular to Σ at every point on the surface, as shown in Figure D.1. The function **n** is the **unit normal** vector field on Σ. Since φ is continuously differentiable, **n** is continuous.

Given such a parametrization, the surface integral of a real-valued, integrable function f defined on a neighborhood of the surface Σ is

$$\int_\Sigma f\mathbf{n} \, da = \int\int_\Omega f(\varphi(\xi_1, \xi_2)) \, \mathbf{n}(\xi_1, \xi_2) \, d\xi_1 \, d\xi_2.$$

Similarly, for an integrable, vector-valued function **f** defined on a neighborhood of Σ,

$$\int_\Sigma \mathbf{f} \cdot \mathbf{n} \, da = \int\int_\Omega \mathbf{f}(\varphi(\xi_1, \xi_2)) \cdot \mathbf{n}(\xi_1, \xi_2) \, d\xi_1 \, d\xi_2.$$

D.2 The Stokes Theorem

The Stokes theorem relates surface integrals of certain derivatives to integrals around the bounding arcs. In this respect, it belongs to a class of theorems—such as the fundamental theorem of calculus and the divergence theorem—that relate integrals of derivatives to values on boundaries. These theorems figure prominently in the mechanics of continua.

In this case, the derivative of interest is the curl

$$\nabla \times \mathbf{F} = \left(\frac{\partial F_3}{\partial x_1} - \frac{\partial F_2}{\partial x_3}, \quad \frac{\partial F_1}{\partial x_3} - \frac{\partial F_3}{\partial x_1}, \quad \frac{\partial F_2}{\partial x_1} - \frac{\partial F_1}{\partial x_2} \right)$$

of a continuously differentiable, vector-valued function $\mathbf{F}(\mathbf{x})$ defined in a region containing a surface Σ that has a well-defined unit normal vector field $\mathbf{n}(\mathbf{x})$.

Let Σ have boundary $\partial\Sigma$, parametrized by a continuously differentiable, closed path γ defined on an interval $[a, b]$ chosen so that the tangent vector field

$\gamma'(s) = \mathbf{t}(s)$ has unit length for every value of the parameter $s \in [a, b]$. The **Stokes theorem** relates the surface integral of $\nabla \times \mathbf{F}$ to the path integral of $\mathbf{F}$:

$$\int_{\Sigma} (\nabla \times \mathbf{F}) \cdot \mathbf{n} \ da = \int_{\gamma} \mathbf{F} \cdot \mathbf{t} \ ds = \int_{a}^{b} \mathbf{F}(\gamma(s)) \cdot \gamma'(s) \ ds. \tag{D.1}$$

See [99, Chapter 8].

D.3 A Corollary to the Stokes Theorem

Section 6.1 uses a corollary of Eq. (D.1) in deriving the Young–Laplace equation (6.2).

Theorem D.1 *If the vector-valued function* $\mathbf{f}$ *is continuously differentiable in a neighborhood of the surface* Σ *and the vector fields* $\mathbf{n}$ *and* $\mathbf{t}$ *are as in Eq. (D.1), then*

$$\int_{\gamma} \mathbf{f} \times \mathbf{t} \ ds = \int_{\Sigma} \left[(\nabla \cdot \mathbf{f}) \mathbf{n} - \mathbf{n} \cdot \nabla \mathbf{f} \right] \ da.$$

Here, the derivative $\nabla \mathbf{f}$ has the following matrix representation with respect to an orthonormal basis $\{\mathbf{e}_1, \mathbf{e}_2, \mathbf{e}_3\}$:

$$\begin{bmatrix} \partial f_1/\partial x_1 & \partial f_1/\partial x_2 & \partial f_1/\partial x_3 \\ \partial f_2/\partial x_1 & \partial f_2/\partial x_2 & \partial f_2/\partial x_3 \\ \partial f_3/\partial x_1 & \partial f_3/\partial x_2 & \partial f_3/\partial x_3 \end{bmatrix}.$$

Using this representation, we calculate the vector $\mathbf{n} \cdot \nabla \mathbf{f}$ via matrix multiplication:

$$\begin{bmatrix} n_1 & n_2 & n_3 \end{bmatrix} \begin{bmatrix} \partial f_1/\partial x_1 & \partial f_1/\partial x_2 & \partial f_1/\partial x_3 \\ \partial f_2/\partial x_1 & \partial f_2/\partial x_2 & \partial f_2/\partial x_3 \\ \partial f_3/\partial x_1 & \partial f_3/\partial x_2 & \partial f_3/\partial x_3 \end{bmatrix}. \tag{D.2}$$

Proof: Since $\mathbf{0}$ is the only vector that is orthogonal to every vector, it suffices to show that the following equation holds for every constant vector $\mathbf{w}$:

$$\mathbf{w} \cdot \left\{ \int_{\gamma} \mathbf{f} \times \mathbf{t} \ ds - \int_{\Sigma} \left[(\nabla \cdot \mathbf{f}) \mathbf{n} - \mathbf{n} \cdot \nabla \mathbf{f} \right] \ da \right\} = 0.$$

Toward this end, let $\mathbf{F} = \mathbf{f} \times \mathbf{w}$ in Eq. (D.1):

$$\int_{\gamma} (\mathbf{f} \times \mathbf{w}) \cdot \mathbf{t} \ ds = \int_{\Sigma} \left[\nabla \times (\mathbf{f} \times \mathbf{w}) \right] \cdot \mathbf{n} \ da.$$

Exercise D.1 *For general, differentiable vector-valued functions* $\mathbf{f}$, $\mathbf{t}$, *and* $\mathbf{w}$, *prove the following vector identity:*

$$(\mathbf{f} \times \mathbf{w}) \cdot \mathbf{t} = -\mathbf{w} \cdot (\mathbf{f} \times \mathbf{t}).$$

If you have exceptional stamina, verify the following additional identity:

$$\nabla \times (\mathbf{f} \times \mathbf{w}) = (\nabla \cdot \mathbf{w})\mathbf{f} - (\nabla \cdot \mathbf{f})\mathbf{w} + (\mathbf{w} \cdot \nabla)\mathbf{f} - (\mathbf{f} \cdot \nabla)\mathbf{w}.$$

Here, the expression $(\mathbf{f} \cdot \nabla)\mathbf{w}$ *has the following representation with respect to the standard orthonormal basis:*

$$\begin{bmatrix} f_1 \dfrac{\partial w_1}{\partial x_1} + f_2 \dfrac{\partial w_1}{\partial x_2} + f_3 \dfrac{\partial w_1}{\partial x_3} \\[2ex] f_1 \dfrac{\partial w_2}{\partial x_1} + f_2 \dfrac{\partial w_2}{\partial x_2} + f_3 \dfrac{\partial w_2}{\partial x_3} \\[2ex] f_1 \dfrac{\partial w_3}{\partial x_1} + f_2 \dfrac{\partial w_3}{\partial x_2} + f_3 \dfrac{\partial w_3}{\partial x_3} \end{bmatrix}.$$

Exercise D.2   *Verify the identity* $[(\mathbf{w} \cdot \nabla)\mathbf{f}] \cdot \mathbf{n} = \mathbf{w} \cdot (\mathbf{n} \cdot \nabla \mathbf{f})$.

Applying the results of Exercises D.1 and D.2 together with the fact that $\nabla \cdot \mathbf{w} = 0$ and $(\mathbf{f} \cdot \nabla)\mathbf{w} = 0$ in our case, we obtain

$$\mathbf{w} \cdot \int_\gamma \mathbf{f} \times \mathbf{t}\, ds = \int_\Sigma \{(\nabla \cdot \mathbf{f})\mathbf{w} \cdot \mathbf{n} - [(\mathbf{w} \cdot \nabla)\mathbf{f}] \cdot \mathbf{n}\}\, da$$

$$= \mathbf{w} \cdot \int_\Sigma \left[(\nabla \cdot \mathbf{f})\mathbf{n} - \mathbf{n} \cdot \nabla \mathbf{f}\right]\, da,$$

as desired. ∎

Bibliography

1 L.M. Abriola, *Multiphase Migration of Organic Compounds in a Porous Medium: A Mathematical Model*, Lecture Notes in Engineering 8, Springer-Verlag, Berlin, 1984.

2 G. Acs, S. Doleschall, and E. Farkas, General purpose compositional model, *Society of Petroleum Engineers Journal* 25:4, 1985, 543–553.

3 A.H. Alizadeh and M. Piri, Three-phase flow in porous media: a review of experimental studies on relative permeability, *Reviews of Geophysics* 52, 2014, 468–521.

4 M.B. Allen, *Continuum Mechanics: Birthplace of Mathematical Models*, John Wiley & Sons, New York, 2016.

5 M.B. Allen and E.L. Isaacson, *Numerical Analysis for Applied Science*, 2nd edition, John Wiley & Sons, Hoboken, NJ, 2019.

6 M.B. Allen and C.L. Murphy, A finite-element collocation method for variably saturated flow in two space dimensions, *Water Resources Research* 22:11, 1986, 1537–1542.

7 L.C. Andrews, *Special Functions for Engineers and Applied Mathematicians*, John Wiley & Sons, New York, 1985.

8 T. Arbogast, J. Douglas Jr., and U. Hornung, Derivation of the double porosity model for single phase flow via homogenization theory, *SIAM Journal of Mathematical Analysis* 21:4, 1990, 823–836.

9 J.-L. Auriault, On the domain of validity of Brinkman's equation, *Transport in Porous Media* 79, 2009, 215–223.

10 A.V. Azevedo, A.J. de Souza, F. Furtado, D. Marchesin, and B. Plohr, The solution by the wave curve method of three-phase flow in virgin reservoirs, *Transport in Porous Media* 83, 2010, 99–125.

11 K. Aziz and A. Settari, *Petroleum Reservoir Simulation*, Elsevier Applied Science Publishers, London, 1979.

12 G.I. Barenblatt, On some unsteady motions of a liquid and gas in a porous medium, *Prikladnaja Matematika i Mekhanika* 16, 67–78, 1952 (Russian).

13 G.I. Barenblatt, V.M. Entov, and V.M. Ryzhik, *Theory of Fluid Flows Through Natural Rocks*, Kluwer, Dordrecht, 1990.

14 G.I. Barenblatt and Ya.B. Zel'dovich, Self-similar solutions as intermediate asymptotics, *Annual Reviews of Fluid Mechanics* 4, 1972, 285–312.

15 G.I. Barenblatt, I.P. Zheltov, and I.N. Kochina, Basic concepts in the theory of seepage of homogeneous liquids in fissured rocks [strata], English translation in *Journal of Applied Mathematics and Mechanics* 24, 1961, 1286–1303.

16 G.K. Batchelor, *An Introduction to Fluid Dynamics*, Cambridge University Press, Cambridge, UK, 1967.

17 J. Bear, On the tensor form of dispersion, *Journal of Geophysical Research* 66, 1961, 1185–1197.

18 J. Bear and A. H.-D. Cheng, *Modeling Groundwater Flow and Contaminant Transport*, Springer, Dordrecht, 2020.

19 P. Bedrikovetsky, *Mathematical Theory of Oil and Gas Recovery With Applications to ex-USSR Oil and Gas Fields*, Kluwer, Dordrecht, 1993.

20 J.B. Bell, J.A. Trangenstein, and G.R. Shubin, Conservation laws of mixed type describing three-phase flow in porous media, *SIAM Journal of Applied Mathematics* 46:6, 1986, 1000–1017.

21 I. Berre, F. Doster, and E. Keilegavlen, Flow in fractured porous media: a review of conceptual models and discretization approaches, *Transport in Porous Media* 130:1, 2019, 215–236.

22 J. Bertrand, Sur l'homogénéité dans les formules de physique, *Comptes Rendus* 86:15, 1878, 916–920.

23 G. Birkhoff, Numerical fluid dynamics, *SIAM Review* 25:1, 1983, 1–34.

24 M.J. Blunt, An empirical model for three-phase relative permeability, *Society of Petroleum Engineers Journal* 5:4, 2000, 435–445.

25 J. Boussinesq, *Essai sur la théorie des eaux courantes, Mémoires présentées par divers savants à l'Académie des Sciences*, Imprimerie Nationale, Paris, 1877.

26 J. Boussinesq, Recherches théoriques sur l'écoulement des nappes d'eau infiltrées dans le sol et sur débit de sources, *Journal de Mathématiques Pures et Appliquées* 10, 1904, 5–78.

27 R.M. Bowen, Incompressible porous media models by use of the theory of mixtures, *International Journal of Engineering Science* 18, 1980, 1129–1148.

28 H.C. Brinkman, A calculation of the viscous force exerted by a flowing fluid on a dense swarm of particles, *Applied Scientific Research* A1, 1947, 27–34.

29 S.E. Buckley and M.C. Leverett, Mechanism of fluid displacement in sands, *Transactions of AIME* 146, 1942, 107–116.

30 E. Buckingham, *Studies on the Movement of Soil Moisture*, Bulletin 38, Bureau of Soils, U.S. Department of Agriculture, Washington, DC, 1907.

31 E. Buckingham, On physically similar systems; illustrations of the use of dimensional equations, *Physical Review IV* 4, 1914, 345–376.

32 M.A. Celia, E.T. Bouloutos, and R.L. Zarba, A general mass conservative numerical solution for the unsaturated flow equation, *Water Resources Research* 26, 1990, 1483–1496.

33 Z. Chen, *Finite Element Methods and Their Applications*, Springer-Verlag, Heidelberg, 2005.

34 Z. Chen and R.E. Ewing, Comparison of various formulations of three-phase flow in porous media, *Journal of Computational Physics* 132, 1997, 362–373.

35 Z. Chen, G. Huan, and Y. Ma, *Computational Methods for Multiphase Flow in Porous Media*, Society for Industrial and Applied Mathematics, Philadelphia, PA, 2006.

36 R.L. Chuoke, P. van Meurs, and C. van der Poel, The instability of slow, immiscible, viscous liquid–liquid displacements in permeable media, *Petroleum Transactions of AIME* 216, 1959, 188–194.

37 F. Civan, C.S. Rai, and C.H. Sondergeld, Shale-gas permeability and diffusivity inferred by improved formulation of relevant retention and transport mechanisms, *Transport in Porous Media* 86, 2011, 925–944.

38 K.H. Coats, An equation of state compositional model, *Society of Petroleum Engineers Journal* 20, 1980, 363–376.

39 H.H. Cooper and C.E. Jacob, A generalized graphical method of evaluating formation constants and summarizing well-field history, *Transactions of the American Geophysical Union* 27:4, 1946, 526–534.

40 A.T. Corey, C.H. Rathjens, J.H. Henderson, and M.R.J. Wyllie, Three-phase relative permeability, *Journal of Petroleum Technology*, 8, 1956, 63–65.

41 G. Dagan, Solute transport in heterogeneous porous formations, *Journal of Fluid Mechanics* 145, 1984, 155–177.

42 G. Dagan, Theory of solute transport by groundwater, *Annual Reviews of Fluid Mechanics* 19, 1987, 183–215.

43 H. Darcy, Les Fontaines Publiques de la Ville de Dijon, Dalmont, Paris, 1856.

44 M. Delshad and G.A. Pope, Comparison of three-phase oil relative permeability models, *Transport in Porous Media* 4, 1989, 59–83.

45 R. di Chiara Roupert, G. Chavent, and G. Schäfer, Three phase compressible flow in porous media: total differential compatible interpolation of relative permeabilities, *Journal of Computational Physics* 229, 2010, 4762–4780.

46 J. Douglas Jr., R.E. Ewing, and M.F. Wheeler, The approximation of the pressure by a mixed method in the simulation of miscible displacement, *RAIRO: Analyse Numérique* 17, 1983, 17–33.

47 J. Douglas Jr., D.W. Peaceman, and H.H. Rachford, A method for calculating multi-dimensional immiscible displacement, *Transactions of the Society of Petroleum Engineers of AIME* 216, 1959, 297–306.

48 F.A.L. Dullien, *Porous Media: Fluid Transport and Pore Structure*, 2nd edition, Academic Press, San Diego, CA, 1992.

49 A.J.E.J. Dupuit, Mémoire sur le mouvement de l'eau à travers les terrains perméables, *Comptes Rendus Hebdomadaire des Séances de l'Académie des Sciences (Paris)* 45, 1857, 92–96.

50 A.J.E.J. Dupuit, *Études Théorique et Practiques sur le Mouvement des Eaux Découverts et à Travers les Terraines Perméables*, 2nd edition, Dunod, Paris, 1863.

51 F.J. Fayers and R.L. Perrine, Mathematical description of detergent flooding in oil reservoirs, *Transactions of AIME* 216, 1959, 277–283.

52 A.E. Fick, On liquid diffusion, *The London, Edinbugh, and Dublin Philosophical Magazine and Journal of Science* 10, 1855, 30–39.

53 P. Forchheimer, Über die Ergiebigkeit von Brunnen-Anlagen und Sickerschlitzen, *Zeitschrift der Architekten- und Ingenieur-Verein* 32, 1886, 539–563.

54 P. Forchheimer, Wasserbewegung durch boden, *Zeitschrift des Vereins Deutscher Ingenieure* 45, 1901, 1782–1788.

55 J.J. Fried and M.A. Combarnous, Dispersion in porous media, in *Advances in Hydroscience*, vol. 7, edited by V.T. Chow, Elsevier, Amsterdam, 169–282, 1971.

56 L.T. Fussell and D.D. Fussell, An iterative technique for compositional reservoir models, *Society of Petroleum Engineers Journal* 1979, 18, 211–220.

57 D.D. Fussell and J.L. Yanosik, (1978) An iterative sequence for phase-equilibria calculation incorporating the Redlich-Kwong equation of state, *Society of Petroleum Engineers Journal*, 19 1978, 173–182.

58 L.W. Gelhar, *Stochastic Subsurface Hydrology*, Prentice Hall, Englewood Cliffs, NJ, 1993.

59 L.W. Gelhar, C. Welty, and K.R. Rehfeldt, A critical review of data on field-scale dispersion in aquifers, *Water Resources Research* 28:7, 1992, 1955–1974.

60 J.W. Gibbs, On the equilibrium of heterogeneous substances, *Transactions of the Connecticut Academy of Arts and Sciences* 3, 1878, 198–248 and 343–524.

61 B.H. Gilding, Qualitative mathematical analysis of the Richards equation, *Transport in Porous Media* 5, 1991, 651–666.

62 J. Glimm, Nonlinear and stochastic phenomena: the grand challenge for partial differential equations, *SIAM Review* 33:4, 1991, 626–643.

63 W.G. Gray and G.F. Pinder, An analysis of the numerical solution of the transport equation, *Water Resources Research* 12:3, 1976, 547–555.

64 R.A. Greenkorn and D.P. Kessler, Dispersion in heterogeneous nonuniform anisotropic porous media, *Industrial and Engineering Chemistry* 61:9, 1969, 14–32.

65 R.B. Guenther and J.W. Lee, *Partial Differential Equations of Mathematical Physics and Integral Equations*, Prentice Hall, Englewood Cliffs, NJ, 1988, republished by Dover, New York, 1996.

66 R.E. Guzman and F.J. Fayers, Solutions to the three-phase Buckley-Leverett problem, *Society of Petroleum Engineers Journal* 37, 1997, 301–311.

67 G.H.L. Hagen, III. Über die Bewgung des Wassers in engen cylindrischen Röhren, *Poggendorfs Annalen der Physik und Chemie (2)* 46, 1839, 423–442.

68 W.B. Haines, Studies in the physical properties of soils. V. The hysteresis effect in capillary properties, and the modes of water distribution associated therewith, *Journal of Agricultural Science* 20:1, 1930, 97–116.

69 S.M. Hassanizadeh and W.G. Gray, General conservation equations for multi-phase systems: 3. Constitutive theory for porous media flow, *Advances in Water Resources* 3, 1980, 25–40.

70 S.M. Hassanizadeh and W.G. Gray, Thermodynamic basis of capillary pressure in porous media, *Water Resources Research* 29:10, 1993, 3389–3405.

71 F.G. Helfferich, Theory of multicomponent, multiphase displacement in porous media, *Society of Petroleum Engineers Journal* 21:1, 1981, 51–62.

72 S. Hill, Channelling in packed columns, *Chemical Engineering Science* 1, 1952, 247–253.

73 G. Homsy, Viscous fingering in porous media, *Annual Review of Fluid Mechanics* 19, 1987, 271–314.

74 G.M. Hornberger, J. Ebert, and I. Remson, Numerical solution of the Boussinesq equation for aquifer-stream interaction, *Water Resources Research* 6:2, 1970, 601–608.

75 M.K. Hubbert, The theory of ground-water motion, *Journal of Geology* 48:8, 1940, 785–944.

76 M.K. Hubbert, Darcy's law and the field equations of the flow of underground fluids, *International Association of Scientific Hydrology Bulletin* 10:1, 1957, 23–59.

77 E.L. Isaacson, D. Marchesin, B.J. Plohr, and B. Temple, Multiphase flow models with singular Riemann problems, *Computational and Applied Mathematics* 11, 1992, 147–166.

78 M.D. Jackson and M.J. Blunt, Elliptic regions and stable solutions for three-phase flow in porous media, *Transport in Porous Media* 48, 2002, 249–269.

79 A.M. Jaffe, The millennium grand challenge in Mathematics, *Notices of the American Mathematical Society*, 53:6, 2000, 652–660.

80 L. Jasinski and M. Dabrowski, The effective transmissivity of a plane-walled fracture with circular cylindrical obstacles, *Journal of Geophysical Research* 123:1, 2018, 242–263.

81 F. Javadpour, D. Fisher, and M. Unsworth, Nanoscale gas flow in shale gas sediments, *Canadian Journal of Petroleum Technology* 46:10, 2007, 55–61.

82 R.T. Johns, B. Dindoruk, and F.M. Orr, Analytical theory of combined condensing/vaporizing gas drives, *Society of Petroleum Engineers Advanced Technology Series*, 1(2), 1993, 7–16.

83 R. Juanes and T.W. Patzek, Analytical solution to the Riemann problem of three-phase flow in porous media, *Transport in Porous Media* 55, 2004, 47–70.

84 R. Juanes and T.W. Patzek, Three-phase displacement theory: an improved description of relative permeabilities, *Society of Petroleum Engineers Journal* 44, 2004, 302–313.

85 J. Jurin, An account of some experiments shown before the Royal Society, with an enquiry into the cause of some of the ascent and suspension of water in capillary tubes, *Philosophical Transactions of the Royal Society of London* 30, 1718, 739–747.

86 M. Karimi-Fard and L. Durlofsky, A general gridding, discretization, and coarsening methodology for modeling flow in porous formations with discrete geological fractures, *Advances in Water Resources* 96, 2016, 354–372.

87 B.L. Keyfitz, Multiphase saturation equations, change of type, and inaccessible regions, in *Proceedings of the Oberwolfach Conference on Porous Media*, edited by J. Douglas Jr., C.J. van Duijn, and U. Hornung, Birkhäuser, Basel, 1993, 103–116.

88 M.J. King, Application and analysis of a new method for calculating tensor permeability, in *New Developments in Improved Oil Recovery*, edited by H.J. deHaan, Geological Society, London, Special Publication 84, 1995, 19–27.

89 M.J. King, P.R. King, C.A. McGill, and J.K. Williams, Effective properties for flow calculations, *Transport in Porous Media* 20, 1995, 169–196.

90 L.J. Klinkenberg, The permeability of porous media to liquids and gases, in *Drilling and Production Practice*, American Petroleum Institute, Washington, DC, 1941, 200–213.

91 L.W. Lake, R.T. Johns, W.R. Rossen, and G.A. Pope, *Fundamentals of Enhanced Oil Recovery*, Society of Petroleum Engineers, Richardson, TX, 2014.

92 I. Langmuir, The adsorption of gases on plane surfaces of glass, mica, and platinum, *Journal of the American Chemical Society* 40:9, 1918, 1361–1403.

93 P.D. Lax, Weak solutions of nonlinear hyperbolic equations and their numerical computation, *Communications on Pure and Applied Mathematics* 7, 1954, 159–193.

94 F.K. Lehner, A derivation of the field equations for slow viscous flow through a porous medium, *Industrial and Engineering Chemistry Fundamentals* 18:1, 1979, 41–45.

95 M.C. Leverett, Flow of oil-water mixtures through unconsolidated sands, *Transactions of the AIME* 132:1, 1939, 149–171.

96 M.C. Leverett, Capillary behavior in porous solids, *Transactions of the AIME* 142:1, 1941, 152–169.

97 M.C. Leverett and W.B. Lewis, Steady flow of gas-oil-water mixtures through unconsolidated sands, *Transactions of the AIME* 142:1, 1941, 107–116.

98 T. Lowry, M.B. Allen, and P.N. Shive, Singularity removal: a refinement of resistivity modeling techniques, *Geophysics* 54:6, 1989, 766–774.

99 J.E. Marsden and A.J. Tromba, *Vector Calculus*, 5th edition, W.H. Freeman & Co., New York, 2003.

100 M.L. Michelsen, The isothermal flash problem. Part I. Stability, *Fluid Phase Equilibria* 9 (1), 1982, 1–19.

101 M.L. Michelsen, The isothermal flash problem. Part II. Phase-split calculation, *Fluid Phase Equilibria* 9 (1), 1982, 21–40.

102 C.T. Miller, G. Christakos, P.T. Imhoff, J.F. McBride, J.A. Pedit, and J.A. Trangenstein, Multiphase flow and transport modeling in heterogeneous porous media: challenges and approaches, *Advances in Water Resources* 21, 1998, 77–120.

103 N.R. Morrow, Physics and thermodynamics of capillary action in porous media, *Industrial and Engineering Chemistry* 62:6, 1970, 32–56.

104 M. Muskat, *The Flow of Homogeneous Fluids Through Porous Media*, J.W. Edwards, Inc., Ann Arbor, MI, 1946.

105 M. Muskat and H.G. Botset, Flow of gas through porous materials, *Physics* 1, 1931, 27–47.

106 M. Muskat and M.W. Meres, The flow of heterogeneous fluids through porous media, *Journal of Applied Physics* 7, 1936, 346–363.

107 M. Muskat, R.D. Wyckof, H.G. Botset, and M.W. Meres, Flow of gas-liquid mixtures through sands, *Transactions of AIME* 123, 1937, 69–96.

108 S.P. Neumann, C.L. Winter, and C.N. Newman, Stochastic theory of field-scale Fickian dispersion in anisotropic porous media, *Water Resources Research* 23:3, 1987, 453–466.

109 L.X. Nghiem, D.K. Fong, and K. Aziz, Compositional modeling with an equation of state, *Society of Petroleum Engineers Journal* 21, 1981, 687–698.

110 P.G. Nutting, Physical analysis of oil sands, *Bulletin of the American Association of Petroleum Geologists* 14, 1930, 1337–1349.

111 O. Oleinik, Discontinuous solutions of nonlinear differential equations, *American Mathematical Society Translation Series* 2:26, 1957, 95–172.

112 L. Onsager, Reciprocal relations in irreversible processes. I, *Physical Review* 37, 1931, 405–426.

113 J.C. Parker, R.J. Lenhard, and T. Kuppasamy, A parametric model for constitutive properties governing multiphase flow in porous media, *Water Resources Research* 23:4, 1987, 618–624.

114 J.-Y. Parlange and E.D. Hill, Theoretical analysis of wetting front instability in soils, *Soil Science* 122, 1976, 236–239.

115 D.W. Peaceman, *Fundamentals of Numerical Reservoir Simulation*, Elsevier Scientific Publishing Company, Amsterdam, 1977.

116 D.-Y. Peng and D.R. Robinson, A new two-constant equation of state, *Industrial and Engineering Chemistry Fundamentals* 15, 1976, 59–64.

117 E.J. Peters and D.L. Flock, The onset of instability during immiscible displacement in porous media, *Society of Petroleum Engineers Journal* 21, 1981, 249–258.

118 J.R. Philip, Flow in porous media, *Annual Reviews of Fluid Mechanics* 2, 1970, 177–204.

119 J.R. Philip, Stability analysis of infiltration, *Soil Science Society of America Proceedings* 39, 1975, 1042–1049.

120 G.F. Pinder and M.A. Celia, *Subsurface Hydrology*, John Wiley & Sons, Hoboken, NJ, 2006.

121 G.F. Pinder and W.G. Gray, *Essentials of Multiphase Flow and Transport in Porous Media*, John Wiley & Sons, Hoboken, NJ, 2008.

122 J.L.M. Poiseuille, Physiques—recherches experimentales sur le mouvement des liquides dans les tubes de tres petits diametres, *Comptes Rendus Academie des Sciences* 11, 1840, 961–967 and 1041–1048.

123 P.Ya. Polubarinova-Kochina, *Theory of Groundwater Movement*, translated from Russian by R.J.M. De Wiest, Princeton University Press, Princeton, NJ, 1962.

124 G.A. Pope, The application of fractional flow theory to enhanced oil recovery, *Society of Petroleum Engineers Journal* 20:3, 1980, 191–205.

125 J.M. Prausnitz, R.N. Lichtenthaler, and E.G. Azevedo, *Molecular Thermodynamics of Fluid-Phase Equilibria*, 3rd edition, Prentice-Hall, Englewood Cliffs, NJ, 1998.

126 O. Redlich and J.N.S. Kwong, On the thermodynamics of solutions, *Chemical Reviews* 44:1, 1949, 233–244.

127 P.C.Y. Raats, Unstable wetting fronts in uniform and non-uniform soils, *Soil Science Society of America Proceedings* 37, 1973, 681–685.

128 O. Reynolds, An experimental investigation of the circumstances which determine whether the motion of water shall be direct or sinuous, and the law of resistance in parallel channels, *Proceedings of the Royal Society of London* 35:224–226, 1883, 84–99.

129 L.A. Richards, Capillary conduction of liquids through porous media, *Physics* 1, 1931, 318–333.

130 A. Robinson, *The Last Man Who Knew Everything: Thomas Young, The Anonymous Polymath Who Proved Newton Wrong, Explained How We See, Cured the Sick, and Deciphered the Rosetta Stone, Among Other Feats of Genius*, Pi Press, New York, 2006.

131 I.F. Roebuck, G.E. Henderson, J. Douglas Jr., and W.T. Ford, The compositional reservoir simulator: case I—the linear model, *Society of Petroleum Engineers Journal* 9 1969, 115–130.

132 P.G. Saffman and G.I. Taylor, The penetration of a fluid into a porous medium or a Hele-Shaw cell containing a more viscous liquid, *Proceedings of the Royal Society A* 245, 1958, 312–239.

133 A.E. Scheidegger, General theory of dispersion in porous media, *Journal of Geophysical Research* 66, 1961, 3273–3278.

134 H. Shahverdi, M. Sohrabi, and M. Jamiolahmady, A new algorithm for estimating three-phase relative permeability from unsteady-state core experiments, *Transport in Porous Media* 90:3, 2011, 911–926.

135 M. Shearer and J.A. Trangenstein, Loss of real characteristics for models of three-phase flow in a porous medium, *Transport in Porous Media* 4, 1989, 499–525.

136 C.S. Slichter, Theoretical investigations of the motion of ground waters, *19th Annual Report, Part II*, U.S. Geological Survey, 1899, 295–384.

137 G. Soave, Equilibrium constants from a modified Redlich-Kwong equation of state, *Chemical Engineering Science* 27:6, 1972, 1197–1203.

138 M. Starnoni and D. Pokrajac, On the concept of macroscopic capillary pressure in two-phase porous media flow, *Advances in Water Resources* 135, 2020, article 103487.

139 G.G. Stokes, On the effect of internal friction of fluids on the motion of pendulums, *Transactions of the Cambridge Philosophical Society II* 9, 1851, 8–106.

140 H.L. Stone, Probability model for estimating three-phase relative permeability, *Journal of Petroleum Technology* 22 1970, 214–218.

141 H.L. Stone, Estimation of three-phase relative permeability and residual oil data, *Journal of Canadian Petroleum Technology* 12(4), 1973, 53–61.

142 H.L. Stone and A.O. Garder Jr., Analysis of gas-cap or dissolved-gas drive reservoirs, *Transactions of the Society of Petroleum Engineers of AIME* 222, 1961, 92–104.

143 D. Swartzendruber, The flow of water in unsaturated soils, in *Flow Through Porous Media*, edited by R.J.M. de Wiest, Academic Press, New York, 1969, 215–292.

144 G.I. Taylor, Dispersion of soluble matter in solvent flowing slowly through a tube, *Proceedings of the Royal Society A* 215, 1953, 186–203.

145 C.V. Theis, The relation between the lowering of the piezometric surface and the rate and duration of discharge of a well using ground water storage, *Transactions of the American Geophysical Union* 16:2, 1935, 519–524.

146 G. Thiem, *Hydrologische Methoden*, Gebhardt, Leipzig, 1906.

147 J.A. Thorpe, *Elementary Topics in Differential Geometry*, Springer, New York, 1979.

148 D.K. Todd, *Groundwater Hydrology*, 2nd edition, John Wiley & Sons, New York, 1980.

149 J.A. Trangenstein and J.B. Bell, Mathematical structure of compositional reservoir simulation, *SIAM Journal of Scientific and Statistical Computing* 10:5, 1989, 817–845.

150 M.Th. van Genuchten, A closed-form equation for predicting the hydraulic conductivity of unsaturated soils, *Soil Science Society of America Journal* 44, 1980, 892–898.

151 J.L. Vásquez, *The Porous Medium Equation: Mathematical Theory*, Oxford University Press, Oxford, 2007.

152 J. von Neumann, First report on the numerical calculation of flow problems, Standard Oil Development Company, 1948, reprinted in *John von Neumann: Collected Works*, vol. 5, edited by A.H. Traub, Macmillan, New York, 1963, 664–750.

153 J. Warren and P. Root, The behavior of naturally fractured reservoirs, *Society of Petroleum Engineers Journal* 3:3, 1963, 245–255.

154 H.G. Welge, A simplified method for computing oil recovery by gas or water drive, *Petroleum Transactions of AIME* 195, 1952, 91–98.

155 R. Wheaton, *Fundamentals of Applied Reservoir Engineering: Appraisal, Economics, and Optimization*, Elsevier, Cambridge, MA, 2016.

156 S. Whitaker, Flow in porous media I: a theoretical derivation of Darcy's law, *Transport in Porous Media* 1, 1986, 3–25.

157 C.D. White and R.N. Horne, Computing absolute transmissibility in the presence of fine-scale heterogeneity, paper SPE 16011, *9th Society of Petroleum Engineers Symposium on Reservoir Simulation*, San Antonio, Texas, February 1–4, 1987, 209–220.

158 Y.-S. Wu, K. Pruess, and P. Persoff, Gas flow in porous media with Klinkenberg effects, *Transport in Porous Media* 32, 1998, 117–137.

159 R.D. Wyckoff and H.G. Botset, The flow of gas-liquid mixtures through unconsolidated sands, *Physics* 7, 1936, 325–245.

160 R.D. Wyckoff, H.G. Botset, M. Muskat, and D.W. Reed, Measurement of permeability of porous media, *Bulletin of the American Association of Petroleum Geologists* 18:2, 1934, 161–190.

161 Y.C. Yortsos and F.J. Hickernell, Linear stability of immiscible displacement in porous media, *SIAM Journal of Applied Mathematics* 49:3, 1989, 730–748.

162 L.C. Young and R.E. Stephenson, A generalized compositional approach for reservoir simulation, *Society of Petroleum Engineers Journal* 23:5, 1983, 727–742.

163 Z. Zeng and R. Grigg, A criterion for non-Darcy flow in porous media, *Transport in Porous Media* 63, 2006, 57–69.

164 Y. Zhou, J.O. Helland, and D.G. Hatzignatiou, Simulation of three-phase capillary pressure curves directly in 2D rock images, paper number IPTC 16547, *International Petroleum Technology Conference*, Beijing, China, March 26–28, 2013.

165 W. Zijl, The symmetry approximation for nonsymmetric permeability tensors and its consequences for mass transport, *Transport in Porous Media* 22, 1996, 121–136.

Index

The Mathematics of Fluid Flow Through Porous Media, First Edition. Myron B. Allen.
© 2021 John Wiley & Sons, Inc. Published 2021 by John Wiley & Sons, Inc.

Printed and bound by CPI Group (UK) Ltd, Croydon, CR0 4YY

16/04/2025

14658343-0002